Constructive Real Analysis

Allen A. Goldstein
Professor Emeritus of Mathematics
University of Washington

Dover Publications, Inc.
Mineola, New York

Bibliographical Note

This Dover edition, first published in 2012, is an unabridged republication of the work originally published in the Harper's Series in Modern Mathematics by Harper & Row, Publishers, New York, in 1967.

Library of Congress Cataloging-in-Publication Data

Goldstein, Allen A.
    Constructive real analysis / Allen A. Goldstein. — Dover ed.
        p. cm.
    Originally published: New York : Harper & Row, 1967; in series: Harper's series in modern mathematics.
    Includes bibliographical references and index.
    ISBN-13: 978-0-486-48879-0
    ISBN-10: 0-486-48879-9
    1. Functions of real variables. I. Title.

QA331.5.G59 2012
515'.8—dc23

                                                    2011042439

Manufactured in the United States by Courier Corporation
48879901
www.doverpublications.com

To L. V. Kantorovich and Victor Klee

CONTENTS

CHAPTER II / CONSTRAINTS

CHAPTER III / INFINITE DIMENSIONAL PROBLEMS

PREFACE

My object in this book is to introduce students of mathematics, science, and technology to the methods of applied functional analysis and applied convexity. To demonstrate and illustrate these techniques I have applied them to the related problems of finding zeros and extremals of functions and operators, with and without side conditions. For students of mathematics the book can perhaps be regarded as an initiation to "hard analysis," while for students of science, an initiation to "soft analysis."

Prerequisite for the reader is a course in advanced calculus (modern style),* linear algebra, and, of course, patience and "elbow grease." The level is upper undergraduate and beginning graduate. The books *Mathematical Analysis* by T. M. Apostol and *Principles of Mathematical Analysis* by W. Rudin will be quoted for references to the calculus. Results from elementary linear algebra will be used without reference. The material can be found for example in the books: *Linear Algebra* by R. Beaumont, or a *Survey of Modern Algebra* by G. Birkhoff and S. Maclane. For supplementary reading in functional analysis, the inexpensive book *Elements of the Theory of Functions and Functional Analysis* by A. N. Komolgoroff and S. V. Fomin is recommended. The reader should observe, however, that certain terms in functional analysis are defined differently in the U.S.S.R. than in the West.

---

* I have, however, avoided the Lesbesque integral.

Many of the problems presented are easy. They are given to aid in the understanding of the text or to elaborate it. I expect the more advanced student to skip over the easy problems, as well as those portions of the text that are already familiar to him.

A preliminary version of this text was prepared at the Massachusetts Institute of Technology from 1961-1963. The book then benefited by repeated teaching at the University of Texas 1963-1964, the University of Washington 1964-1966, and the University of Hamburg 1966-1967.

I am grateful to many people for suggestions with this manuscript, notably R. De Marr (and his students), K. Fan, V. Klee, B. Kripke, and J. Ryff who meticulously read Chapter I and made many improvements. It is a pleasure to thank Marian Goldstein and Victor Klee for their encouragement. To my collaborators in research on topics related to this book, particularly Ward Cheney, I offer my heartfelt thanks. To my students who have found errors in the past and who will continue to do so in the future, I again offer thanks, especially to H. Chiang and F. Johnson. Finally, special thanks are reserved for Susan Moser for typing the final manuscript and to Martha Goldstein for secretarial help. Acknowledgement is also due to the Air Force Office of Scientific Research and to the University of Washington for support during various phases of the work on the manuscript.

Since the main purpose of this book is to stimulate interest in applied functional analysis and applied convexity, it is fitting that this book be dedicated to L. V. Kantorovich and Victor Klee.

A. A. G.

Hamburg, Germany
December, 1966

Constructive
Real Analysis

CHAPTER I

ROOTS AND EXTREMAL PROBLEMS

INTRODUCTION

This book is concerned with processes for finding roots of functions. Often the function f whose root we seek is the derivative of some other function g, that is, $g' = f$. In this case, it can happen that a root of f is an *extremal* (maximum or minimum) for g. These matters, and generalizations of them, will be considered below. We shall dwell on the problem of finding extremals of functions that are subject to restraint.

By an *algorithm* for a problem we mean a recipe by means of which the solution to the problem can be calculated. An *effective algorithm* is one that can be used without exorbitant cost on a modern automatic digital calculator. We shall discuss below algorithms that are not necessarily effective. Sometimes these algorithms are effective for some problems but not for others. Moreover, in many cases, no estimates of effectiveness are known. As a general method the contraction mapping theorem and other principles will be used to discuss algorithms. The heart of the book, however, will be the techniques of the detailed analysis of the algorithms presented. An example for which it is easy to give a detailed analysis follows.

We are concerned with finding the square root of a positive number by a simple iterative computation. The iteration formula is given by $x_{n+1} = \frac{1}{2}(x_n + a/x_n)$. Here, a is a positive number whose square root we wish to extract. The above formula is a consequence of Newton's method, Section A-2. We shall describe the sequence $\{x_n : n = 1, 2, 3, \cdots\}$, which this formula generates. If x_n approximates $\sqrt{a}$, recall that the *relative error* of this approximation is given by

$$\left| \frac{x_n - \sqrt{a}}{\sqrt{a}} \right|.$$

Theorem

(i) Assume a and x_0 to be positive.
(ii) Define the sequence $\{x_n\}$ by $x_{n+1} = \frac{1}{2}(x_n + a/x_n)$.
(iii) Let $\delta_n = (x_n - \sqrt{a})/\sqrt{a}$.
Then

(a) $\delta_{n+1} = \dfrac{1}{2} \dfrac{\delta_n^2}{1 + \delta_n}$, $n = 0, 1, 2, \cdots$;

(b) $\delta_n \geq 0$, $n = 1, 2, 3, \cdots$;

(c) for every $\varepsilon > 0$,

$$\frac{x_n}{a}(x_n - x_{n+1}) < \varepsilon$$

implies $|\delta_n| < \varepsilon$, for $n = 1, 2, 3, \cdots$.

Proof

By (iii), $x_n = \sqrt{a}(\delta_n + 1)$. Using (ii),

$$x_{n+1} = \frac{1}{2}\left(\sqrt{a}(\delta_n + 1) + \frac{a}{\sqrt{a}(\delta_n + 1)} \right) = \sqrt{a}\left[1 + \frac{1}{2}\frac{\delta_n^2}{(1 + \delta_n)} \right].$$

By (iii), $x_{n+1} = \sqrt{a}(1 + \delta_{n+1})$. Hence (a) is proved.

To prove (b), observe that $x_0 = \sqrt{a}(\delta_0 + 1)$, and by (i), $\delta_0 + 1$ is positive. Since δ_0^2 is nonnegative, δ_1 is nonnegative by (a). Thus, (b) follows by use of (a) and induction.

To prove (c), we calculate that

$$\frac{x_n}{a}(x_n - x_{n+1}) = \frac{1}{2}\left(\frac{x_n^2 - a}{a} \right),$$

using (ii). By (c),

$$\frac{x_n^2 - a}{a} < 2\varepsilon < 2\varepsilon + \varepsilon^2.$$

Thus, $x_n^2 < a(1 + 2\varepsilon + \varepsilon^2)$, and $x_n < \sqrt{a}(1 + \varepsilon)$. Hence

$$\frac{x_n - \sqrt{a}}{\sqrt{a}} = \delta_n = |\delta_n| < \varepsilon \qquad \text{for } n = 1, 2, 3,$$

by (b). Q.E.D.

Of course the above analysis does not consider that, in practice, computations will be made with inexact numbers. Considerations of such effects are referred to Wilkinson.[1]

Based on the above theorem, a possible way of calculating $\sqrt{a}$ would be as follows: Given a, x_0, and ε, we calculate the sequence $x_1, x_2, \cdots$ by (ii). For $n \geq 1$, we calculate

$$\frac{x_n - x_{n+1}}{a/x_n} = q_n.$$

When $q_n \leq \varepsilon$, we terminate, for then $|\delta_n| \leq \varepsilon$. In this scheme the termination of computation is done with a posteriori information—that is, the termination depends on the sequence $x_1, x_2, \cdots$. A different algorithm could be stated in which the termination is a priori. The algorithm would be constructed to utilize (a), and would predict the number of iterations required to obtain the $\sqrt{a}$ to desired accuracy.

Unfortunately, we do not usually know as much about an algorithm as we know in this case. The above description is therefore in the nature of an ideal that we strive toward in our study of iterative processes.

PROBLEMS

1. Using (a) and (b) prove that $\{\delta_n\}$ monotonically converges to 0 and hence $\{x_n\}$ converges to $\sqrt{a}$.

2. The inequality $0 \leq \delta_{n+1} \leq \frac{1}{2}\delta_n^2$, $(n = 1, 2, 3, \cdots)$, a consequence of (a) and (b) above, implies $\delta_{n+1} \leq (\frac{1}{2})^{2^n - 1}\delta_1^{2^n}$. Prove this by induction and show that $2^n \log(\delta_1/2) \leq \log \varepsilon/2$ implies that $0 \leq \delta_{n+1} \leq \varepsilon$.

3. Estimate the number of iterations necessary to calculate $\sqrt{3}$ with a relative error of 10^{-10}. Take $x_0 = 1$.

---

[1] J. H. Wilkinson, *Rounding Errors in Algebraic Processes*, Prentice-Hall, Englewood Cliffs, N.J., 1963.

SECTION A. ITERATIONS AND FIXED POINTS

A-1. Functional Iteration and Roots: One Variable

Let $I = [a, b]$ and assume that f is a continuous function on I, taking values on a closed interval of the reals, say J. The function f is also called a *map* from I into J. If J is a subset of I, then f is called a map from I *into itself.* Suppose f is a map from I into itself. A *fixed point* of f is a point $x \in I$ such that $f(x) = x$.

Lemma 1

Every continuous function f mapping I into itself has a fixed point.

Proof

If a is not a fixed point, then $f(a) > a$. If b is not a fixed point, then $f(b) < b$. Let $h(x) = f(x) - x$. Then $h(a) = f(a) - a > 0$ and $h(b) = f(b) - b < 0$. Since h is continuous, h has a root in I at, say, z. Consequently, $f(z) = z$.

A map from I into itself is said to be a *contractor* if, for every pair of points x and y in I, $|f(x) - f(y)| \le q|x - y|$, with $q < 1$.

Theorem 2

Assume that f is a contractor on I and set $x_{n+1} = f(x_n)$ with x_0 in I. Then there exists a unique fixed point of f, say z; the sequence $\{x_n\}$ converges to z; and $|x_{n+1} - z| \le q^{n+1}|x_0 - z|$.

Proof

By the lemma, the existence of a fixed point is ensured. Hence, $|x_{n+1} - z| = |f(x_n) - f(z)| \le q|x_n - z|$. This establishes the formula

$$|x_{n+1} - z| \le q^{n+1}|x_0 - z|$$

for $n = 0$. Assume that $|x_n - z| \le q^n|x_0 - z|$ for any $n \ge 1$. Then

$$|x_{n+1} - z| \le q|x_n - z| \le q^{n+1}|x_0 - z|.$$

Because $q < 1$, $\{x_n\} \to z$. To prove that z is unique, suppose there were two distinct fixed points z_1 and z_2. Then

$$0 < |z_1 - z_2| = |f(z_1) - f(z_2)| \le q|z_1 - z_2| < |z_1 - z_2|,$$

a contradiction.

The process of generating a sequence by the formula $x_{n+1} = f(x_n)$ is often called *iteration* or *functional iteration*. This is because $x_1 = f(x_0)$, $x_2 = f(x_1) = f(f(x_0)) \cdots$, and $x_n = f_n(x_0)$. Here f_n stands for the composition of f n-times. In this terminology we have proved that $\{f_n(x_0)\} \to z$.

Our object now is to determine suitable hypotheses (that is, hypotheses that can be verified and sometimes fulfilled) on a function h in order to calculate its roots by iteration.

Lemma 2

If f has a continuous derivative on I and if f maps I into itself, then $|f'(x)| < 1$ on I implies that f is a contractor.

Proof

By the mean-value theorem, $|f(x) - f(y)| = |f'(\xi)|\,|x - y|$, with ξ between x and y. But $\max_{\xi \in I}|f'(\xi)|$ is less than unity; hence, f is a contractor.

Assume that h is a monotone function on I, which has a continuous positive derivative. Suppose h has a root z in the interior of I. Then $h(a) < 0 < h(b)$. Define $F(x) = x - \lambda h(x)$. If F is to be a map of I into itself, we must have $a \le F(x) \le b$ for all $x \in I$. Observe that if $\lambda > 0$, $F(a) > a$ and $F(b) < b$. Thus, one suspects that for λ small and positive, $a < F(x) < b$ for all x in I. Furthermore, since $F'(x) = 1 - \lambda h'(x)$, and $h'(x) > 0$, it also is evident that for λ small and positive, $|F'(x)| < 1$ on I. These ideas are stated formally in the next paragraph.

Theorem 3

Assume h belongs to $C^1[a, b]$, $h(a)h(b) < 0$, and $0 < \mu \le h'(x) \le 1/\gamma$ for all x in $[a, b]$. Set $x_{n+1} = x_n - \gamma h(x_n)$ with x_0 arbitrary in $[a, b]$. Then $\{x_n\}$ converges to a root z of h and $|x_{n+1} - z| \le (1 - \mu\gamma)^{n+1}|x_0 - z|$.

Proof

Set $F(x) = x - \gamma h(x)$. Observe that z is a fixed point of F if and only if z is a root of h. Since for all x in $[a, b]$, $0 < \mu\gamma \le \gamma h'(x) \le 1$, it follows that $0 \le 1 - \gamma h'(x) \le 1 - \mu\gamma < 1$. Thus, $0 \le F'(x) \le 1 - \mu\gamma < 1$ for all x in $[a, b]$, and F is monotone-increasing on $[a, b]$. Since $h(a)h(b) < 0$ and h is monotone-increasing, $h(a) < 0$. Hence, $F(a) > a$, and $F(b) < b$ because $\gamma > 0$. Because F

is monotone, $a < F(x) < b$ for all x in $[a, b]$. Furthermore, $|F'(x)| \le 1 - \mu\gamma < 1$ for all x in $[a, b]$. Thus, the theorem follows by Theorem 2 and Lemma 2.

Observe that the root z is unique because F has a unique fixed point. If h is monotone-decreasing, $-h$ is monotone-increasing and has the same roots as h.

PROBLEM

The formula $|x_{n+1} - z| \le (1 - \mu\gamma)^{n+1}|x_0 - z|$ would be of practical interest if we could estimate $|x_0 - z|$, since $1 - \mu\gamma$ can be estimated. Show that

$$\left|\frac{h(x_n)}{\mu}\right| \ge |x_n - z| \ge \gamma|h(x_n)|, \qquad n = 0, 1, 2, \cdots$$

We conclude with the observation that Theorem 3 above assures us we can, roughly speaking, calculate roots, provided we can isolate an interval in which a function is monotone and changes sign.

It should be emphasized that the conditions of the above theorem are sufficient but not necessary for the application of the fixed-point theorem. For example, the hypothesis that μ is positive can be relaxed. This will be done in Section A-3. In Section A-2, we cite an example in which $F'(x)$ changes sign on $[a, b]$.

Example

Let $h(x) = x^2 - \alpha$, and assume $a > 0$. Suppose further that $\alpha \in (a^2, b^2)$. Then $h(a) < 0$ and $h(b) > 0$. Also, $0 < 2a \le h'(x) \le 2b$ for all x in $[a, b]$. By Theorem 3 above, the sequence

$$\{x_{n+1}\} = \left\{x_n - \frac{1}{2b}(x_n^2 - \alpha)\right\}$$

converges to $\sqrt{\alpha}$ whenever $x_0 \in [a, b]$. The rate of convergence is given by

$$|x_{n+1} - \sqrt{\alpha}| \le \left(1 - \frac{a}{b}\right)^{n+1}|x_0 - \sqrt{\alpha}|.$$

In Theorem 3, other iterations could be defined. For example, we could have set $x_{n+1} = x_n - \lambda h(x_n)$ with $0 < \lambda \le \gamma$. Indeed, it is not even necessary that λ be chosen a constant. For example, let λ be a function in $C^1[a, b]$ such that $\lambda(x)$ is positive on $[a, b]$ and, as in Theorem 3, λ is chosen in such a way that for $x \in [a, b]$, $0 \le F'(x) = 1 - \lambda'(x)h(x) - h'(x)\lambda(x) < 1$, where $F(x) = x - \lambda(x)h(x)$.

A-2. Newton's Method

Set $\lambda(x) = 1/h'(x)$ and assume the hypotheses on h of Theorem 3. Assume moreover, $h \in C^2[a, b]$, and calculate that

$$F'(x) = \frac{h''(x)h(x)}{(h'(x))^2}.$$

The iteration defined by

$$x_{n+1} = F(x_n) = x_n - \frac{h(x_n)}{h'(x_n)}$$

is called *Newton's method*. By Theorem 2 and Lemma 2, we obtain convergence of the sequence $\{x_n\}$, provided $|F'(x)| \le q < 1$ for all $x \in [a, b]$, and F is a map of $[a, b]$ into itself.

Newton's method has the following geometrical motivation: Given x_n, approximate h by the tangent line at $(x_n, h(x_n))$. Choose x_{n+1} to be the zero crossing of this tangent line. Let $\bar{h}$ denote the linear function that approximates h in this way. Then $\bar{h}(x_{n+1}) = 0 = h(x_n) + h'(x_n)(x_{n+1} - x_n)$.

A pleasant feature of Newton's method is that if the sequence generated by it converges, the rate at which convergence takes place is quadratic. To see this, assume that z is a fixed point of F and write $x_n - z = F(x_{n-1}) - F(z) = F'(\xi_n)(x_{n-1} - z)$ or

$$|x_n - z| = \frac{|h''(\xi_n)h(\xi_n)|}{h'^2(\xi_n)} |x_{n-1} - z|,$$

where ξ_n is between x_{n-1} and z. Assume $\{x_n\} \to z$. Expanding h around its root, we get

$$|h(\xi_n) - h(z)| = |h(\xi_n)| = |h'(\eta)| \, |\xi_n - z| < |(h'(\eta_n)| \, |x_{n-1} - z|,$$

η_n between ξ_n and z. Put $B_n = |h''(\xi_n)h'(\eta_n)/h'^2(\xi_n)|$ and $B = \sup_n B_n$. Then $|x_n - z| < B|x_{n-1} - z|^2$. Observe that as $n \to \infty$, $B_n \to |h''(z)/h'(z)|$. A detailed study of Newton's method in a more general setting will be found in Chapter III, Section C-4.

Example

If Newton's method is applied to the function $x^2 - \alpha$, we obtain the familiar formula

$$x_{n+1} = \frac{1}{2}\left(x_n + \frac{\alpha}{x_n}\right).$$

We shall show that, given α, an interval $[a, b]$ exists such that the function F of Newton's method is a contractor. Choose a and b such that $0 < a^2 < \alpha < b^2$, $b > (a^2 + \alpha)/2a$ and $3a^2 > \alpha$. For example, $a = \alpha/2$ and $b > 3a$.[†] We shall verify that $x \in [a, b]$ implies $a \le F(x) = (x^2 + \alpha)/2x \le b$ by showing that the minimum and maximum values of F on $[a, b]$ belong to $[a, b]$.

Calculate that $F'(x) = \frac{1}{2}(1 - (\alpha/x^3))$ and $F''(x) = \alpha/x^3 > 0$. Thus, $F'(\sqrt{\alpha}) = 0$ and $F(\sqrt{\alpha}) = \alpha \in [a, b]$. Evidently, the maximum of F must occur at the end points a or b because F' vanishes only once. But $F(a) = (a^2 + \alpha)/2a < b$ and $F(b) = (b^2 + \alpha)/2b < b$. To conclude the proof, we calculate that $-\alpha/a^2 > -3$ implies that

$$-1 < \tfrac{1}{2}(1 - (\alpha/a^2)) \le \tfrac{1}{2}(1 - (\alpha/x^2)) \le \tfrac{1}{2}(1 - \alpha/b^2) < \tfrac{1}{2}.$$

Thus, $|F'(x)| < 1$ on $[a, b]$. Observe that an easy choice for x_0 is $x_0 = \alpha + 1 > \sqrt{\alpha}$, since if $a = \sqrt{\alpha/2}$ and $b = \max\{3a, \alpha + 1\}$, $x_0 \in [a, b]$.

A-3. Subcontractors

A *subcontractor* is a map f of a finite interval I into itself such that

(i) x and y in I imply $|f(x) - f(y)| \le |x - y|$,
(ii) if $x \ne f(x)$, $|f(f(x)) - f(x)| < |f(x) - x|$.

Theorem

Assume f is a subcontractor on I. Choose x_0 arbitrarily in I and set $x_{n+1} = f(x_n)$. Then $\{x_n\}$ converges to a fixed point of f.

Proof

Deferred to Section B-5.

Lemma

Assume $f \in C^1[a, b]$, $0 \le f'(x) \le 1$, and $f'(x) \ne 1$ for some $x \in [a, b]$. Then

$$0 \le \frac{1}{b - a} \int_a^b f'(t) \, dt < 1.$$

---

† We assume $\frac{4}{3} < \alpha < 4$. Observe that $3a = \dfrac{3a^3}{a^2} > \dfrac{a\alpha}{a^2} = \dfrac{2\alpha}{2a} > \dfrac{a^2 + \alpha}{2a}$.

Proof

Since f' is continuous on $[a, b]$, it has a minimum on $[a, b] = I$, say at z. Then $f'(z) = q < 1$. There exists an open interval N in $[a, b]$ such that $x \in N$ implies $f'(x) < (1 + q)/2$. Let $\mu(N)$ denote the length of the interval N. Then

$$\int_a^b f'(t)\, dt = \int_N f'(t)\, dt + \int_{I \sim N} f'(t)\, dt < \mu(N)\frac{1 + q}{2} + (b - a) - \mu(N)$$

$$= \mu(N)\left[\frac{1 + q}{2} - 1\right] + (b - a) < b - a.$$

Corollary

Assume that h belongs to $C^1[a, b]$, $h(a)h(b) < 0$, $0 \leq \mu \leq h'(x) \leq 1/\gamma$, and for each subinterval I' of $[a, b]$, there exists $x \in I'$ such that $h'(x) \neq 0$. Set $x_{n+1} = x_n - \gamma h(x_n)$ with x_0 arbitrary in $[a, b]$. Then $\{x_n\}$ converges to a root of h.

PROBLEM

Show by example that h' may have infinitely many zeros on $[a, b]$, and satisfy the above hypotheses.

Proof of Corollary

Reasoning as in Section A-1, we have for all x in $[a, b]$ that

$$0 \leq F'(x) \leq 1 - \gamma h'(x) \leq 1,$$

and $a < F(x) < b$. Also, $|F(x) - F(y)| \leq |x - y|$ for all x and y in $[a, b]$. Choose x_0 in I. If x_0 is a root, we are done; otherwise, $F(x_0) \neq x_0$. Then

$$|x_2 - x_1| = |F(F(x_0)) - F(x_0)| = |F(x_1) - F(x_0)|$$

$$= |x_1 - x_0|\left|\left[\frac{1}{x_1 - x_0}\int_{x_0}^{x_1} F'(t)\, dt\right]\right| < |x_1 - x_0| = |F(x_0) - x_0|,$$

by the lemma above. Hence, by the above theorem, $\{x_n\} \to z$ such that $h(z) = 0$.

Observe that for subcontractors, we have no information on rates of convergence.

We remark that the hypotheses for each subinterval I', $f'(x) \neq 0$ for some $x \in I'$.

I' can be replaced by the hypotheses that f has a power series representation on $[a, b]$ and $f' \not\equiv 0$. Then it follows that f' can vanish only on a countable set of points, showing that for every I', $f'(x) \neq 0$ for some $x \in I'$.

Example

Let $h(x) = \frac{1}{2}x - \frac{1}{4}\sin 2x$, with $[a, b] = [-n\pi, n\pi]$, n a positive integer. The derivative h' vanishes at $j\pi$, $j = 0, \pm 1, \cdots, \pm n$ because $h'(x) = \sin^2 x$. The function h is monotone on $[a, b]$ and has a root at $x = 0$. The example therefore satisfies the hypotheses of the preceding corollary.

SECTION B. METRIC SPACES

B-1. Definitions

Metric spaces originated from the idea of the limiting operations used in analysis. A metric space is a pair consisting of an arbitrary set (say, M) and a nonnegative real valued function (say, d) defined on ordered pairs of elements from M, that is, on $M \times M$. The metric space may be denoted by (M, d) or, when the function d is well fixed in mind, by M.

The function d is assumed to satisfy the following postulates:

(i) $d(x, y) = 0$ if and only if $x = y$;

(ii) $d(x, y) = d(y, x)$;

(iii) $d(x, y) + d(y, z) \geq d(x, z)$ for all x, y, z in M.

The real numbers with $d(x, y) = |x - y|$ constitute a metric space. We readily verify that $|x - y|$ satisfies the above relations.

PROBLEM

Assume postulate (i); assume $d(x, y) \leq d(z, x) + d(z, y)$. Prove that $d(x, y) \geq 0$ and $d(x, y) = d(y, x)$.

Another example of a metric space is the space $C[a, b]$ of continuous functions on the closed interval $[a, b]$. We must keep in mind that the points of M are the functions themselves and not their values. For the metric, we put

$$d(f, g) = \max_{t \in [a, b]} |f(t) - g(t)|.$$

Clearly, axioms (i) and (ii) are satisfied. We have by the triangle inequality for real numbers that $|f(t) - h(t)| + |h(t) - g(t)| \geq |f(t) - g(t)|$ for all t. Thus

$$\max|f(t) - g(t)| \leq \max\{|f(t) - h(t)| + |h(t) - g(t)|\}$$
$$= |f(t_0) - h(t_0)| + |h(t_0) - g(t_0)| \qquad \text{for some } t_0 \in [a, b].$$

But

$$|f(t_0) - h(t_0)| \leq \max|f(t) - h(t)| \quad \text{and} \quad |h(t_0) - g(t_0)| \leq \max|h(t) - g(t)|.$$

Hence, $\max|f(t) - g(t)| \leq \max|f(t) - h(t)| + \max|h(t) - g(t)|$, proving (iii).

A familiar example is the space E_n, which is the collection of real n-tuples; thus, if $x \in E_n$, $x = (x_1, \cdots, x_n)$, x_i real, $(i = 1, \cdots, n)$. In the space E_n we define the metric

$$d(x, y) = \left(\sum_{i=1}^{n} (x_i - y_i)^2 \right)^{1/2} = \|x - y\|.$$

Using the *inner product*

$$[x, y] = \sum_{i=1}^{n} x_i y_i ,$$

we can write this as $d(x, y) = [x - y, x - y]^{1/2}$. Clearly, (i) and (ii) hold. To prove (iii), we must show that $\|x - y\| + \|y - z\| \geq \|x - z\|$. Put $\xi = x - y$ and $\eta = y - z$; then $\xi + \eta = x - z$. Thus, if for every ξ and η in E_n, $\|\xi + \eta\| \leq \|\xi\| + \|\eta\|$, (iii) will hold. We recall that this is the *triangle inequality* for n-tuples.[2]

PROBLEM

In E_n define

$$d(x, y) = \sum_{i=1}^{n} |x_i - y_i|.$$

Show that (E_n, d) is a metric space.

B-2. Review

In the plane (E_2), two points x and y are "close" if they lie in some circle with "small" radius. Let $N_\varepsilon(z) = \{x \in E_2 : (x_1 - z_1)^2 + (x_2 - z_2)^2 < \varepsilon^2\}$ be a disc of radius ε about z. A sequence $\{x_n\}$ *converges to* x if, given $\varepsilon > 0$,

[2] See T. M. Apostol, *Mathematical Analysis*, Addison-Wesley Publishing Co., Reading, Mass., 1957, p. 47, and W. Rudin, *Principles of Mathematical Analysis*, McGraw-Hill, New York, 1953, p. 18.

there exists $N > 0$ such that *for all* $n \geq N$, x_n is in $N_\varepsilon(x)$. The point x is said to be the *limit point* for the sequence $\{x_n\}$. If, given $\varepsilon > 0$, there exists $N > 0$ such that *for infinitely many* $n \geq N$, x_n is in $N_\varepsilon(x)$, then x is said to be an *accumulation point* or *cluster point* for the sequence $\{x_n\}$. If x is a cluster point for the sequence $\{x_n\}$, then a subsequence $\{x_{n_k}\}$ can be constructed converging to x. To do this, let $n_1 = $ smallest n such that $x_n \in N_1(x)$; let $n_2 = $ smallest $n \geq n_1$ such that $x_n \in N_{1/2}(x)$; $n_k = $ smallest $n \geq n_{k-1}$ such that $x_n \in N_{1/k}(x)$. Thus, the subsequence $\{x_{n_k}\}$ is a sequence converging to x. If $x \in E_2$ and for every $\varepsilon > 0$, $N_\varepsilon(x)$ contains infinitely many points of a subset S of E_2, then x is said to be an *accumulation point for S*. A set S in E_2 is said to be *closed* if and only if it contains all its accumulation points. The *closure* of a set S is S together with all its accumulation points. Thus, if S is closed and $\{x_k\} \in S$ is a sequence converging to x, then x belongs to S. An *interior point* x of a set S has the property that for some $\varepsilon > 0$, $N_\varepsilon(x)$ is a subset of S. An *open set* is a set such that each of its points is an interior point. A *neighborhood* of a point x is an open set containing x, for example $N_\varepsilon(x)$.

A necessary and sufficient condition that a sequence converge is that the sequence be a *Cauchy sequence*. A sequence is Cauchy if, given $\varepsilon > 0$, there exists N such that n and $m \geq N$ imply $x_m \in N_\varepsilon(x_n)$. That every Cauchy sequence has a limit follows because the "Cauchyness" implies that the sequence is bounded and hence can be taken to lie in a closed-bounded set. By the *Bolzano-Weierstrass theorem*, the sequence has an accumulation point. The "Cauchyness" implies that this accumulation point is a limit point.[3]

B-3. More Definitions and Information from Analysis

In a metric space the Bolzano-Weierstrass theorem is not usually valid, and hence a bounded sequence (that is, a sequence satisfying $d(x_n, z) \leq B$) does not necessarily have an accumulation point. (See example below.) To generalize arguments in discussions that use the Bolzano-Weierstrass theorem, two convenient hypotheses have been formulated. The first is *completeness*. A *complete metric space* is a space in which every Cauchy sequence has its limit. The second is *compactness*. An *open cover* for a set S is a family of open sets whose union contains S. A subset S of a metric space is called *compact* if every open cover for S contains a finite subcover. If S is the whole metric space and S is compact, S is called a *compact metric space*. If S is compact every sequence in S has a converging subsequence.[4] If S is a subset of a metric space with the property that every infinite sequence has a converging

---

[3] See Apostol, *op. cit.*, Chaps. 2, 3, and 4; and Rudin, *op. cit.*, Chaps 2 and 3.
[4] Apostol, *op. cit.*, p. 55; Rudin, *op. cit.*, p. 33.

subsequence, then S is compact.[5] Compact subsets of a metric space are closed, and closed subsets of compact sets are compact.[6]

Let F be a map (synonymously, operator or function) from a subset S of a metric space M_1 to a metric space M_2. F is called *continuous* at x in S if, given $\varepsilon > 0$, there exists $\delta > 0$ such that when $z \in S$ satisfies $d_1(z, x) < \delta$, then Fz satisfies $d_2(Fz, Fx) < \varepsilon$. Here d_i indicates the metric in M_i, $(i = 1, 2)$.

F is *uniformly continuous* on S if, given $\varepsilon > 0$, there exists $\delta > 0$ such that $d_1(z, x) < \delta$ implies $d_2(Fz, Fx) < \varepsilon$ for all $x \in S$. Stated otherwise, if F is uniformly continuous on S, δ is a function only of ε and not of x. If S is compact and F is continuous on S (that is, F is continuous at every point of S), then F is uniformly continuous.[7] A continuous function on a compact set achieves its maximum and minimum.[8]

Compact subsets of metric spaces that are not closed-bounded subsets of Euclidean n-space are not often met in applications because such sets are "scarce." For an example, we refer the reader to Arzela's theorem, which characterizes some compact subsets of the metric space $C[a, b]$.[9]

A subset S of a metric space M is said to be *dense* or *dense in M* if the closure of S (written $\bar{S}$) is M. A metric space is *separable* if it contains a dense denumerable subset.

Example

Closed-bounded subsets of $C[a, b]$ are not necessarily compact.
Let $S = \{z \in C[0, 1]: z(0) = 0, z(1) = 1, \text{ and } \|z\| = 1\}$. Here

$$\|z\| = \max\{|z(t)|: 0 \le t \le 1\}.$$

Clearly, S is bounded. To see that S is closed, take a sequence $\{x_k\}$ of points from S which is converging to a point x. (This means $\{\|x_k - x\|\} \to 0$.) Observe that we have at hand an example of uniform convergence.[10] But if $\{x_n\}$ is a sequence of continuous functions converging uniformly to x, then x is continuous.[11] Clearly, $x(0) = 0$ and $x(1) = 1$, for to suppose otherwise would be contradictory. It follows that S is closed.

Define the function

$$z \to f(z) = \int_0^1 z^2(t)\, dt.$$

---

[5] Rudin, *op. cit.*, p. 40, Problem 13.

[6] *Ibid.*, p. 33.

[7] Apostol, *op. cit.*, p. 74; Rudin, *op. cit.*, p. 78.

[8] Apostol, *op. cit.*, p. 73; Rudin, *op. cit.*, p. 77.

[9] A. N. Kolmogoroff and S. V. Fomin, *Elements of the Theory of Functions and Functional Analysis*, Graylock Press, Rochester, N.Y., 1957, p. 54.

[10] Apostol, *op. cit.*, p. 393; Rudin, *op. cit.*, p. 133.

[11] Apostol, *op. cit.*, p. 394; Rudin, *op. cit.*, p. 136.

The function f is continuous and positive on S. The positivity follows because $z(1) = 1$ and therefore, by continuity, there exists a neighborhood containing 1 in which $z(t) > \frac{1}{2}$. Let $z = t^n$. Then $f(z) = 1/(2n + 1)$ for every n. Thus, $\inf\{ f(z): z \in S\} = 0$. But this infimum is never achieved. Therefore, S is not compact.

E X E R C I S E S

    1. Prove that f in the above example is continuous on S.

    2. A tent function in $C[0, 1]$ is defined as follows: Let $I = [a, b]$ with $0 \le a < b \le 1$. A tent function $x \to T(I, x)$ has the following graph:

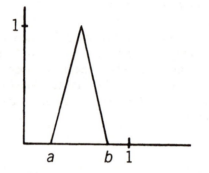

The value of T is 0 on $[0, a]$, linear on $[a, (a + b)/2]$ with

$$T(I, (a + b/2)) = 1,$$

linear on $[(a + b)/2, b]$, and 0 on $[b, 1]$. Exhibit a sequence of intervals $\{I_n\}$ such that the corresponding sequence of x tent functions has no converging subsequence.

Example: Rationals and Reals

    Let E denote the set of all rational numbers. E is a metric space under the metric $d(x, y) = |x - y|$. E is not complete because the Cauchy sequence $\{(1 + 1/n)^n\}$, which converges to the number e, has no limit in E. The space F of reals, which can be obtained, roughly speaking, by augmenting E by the limits of all possible Cauchy sequences of rationals, is complete.[12] Since E is denumerable and dense in F (the closure of E in the metric space F is F itself), F is separable. The distinction between "closed" and "complete" should not be confused. Thus, every metric space is closed, but not every metric space is complete. For example, the rationals are closed in themselves even though they are not closed in the reals.

---

[12] Specifically, see Rudin, p. 71, Exercises 18, 19, and 20.

Space of Polynomials

The metric space of all polynomials on $[a, b]$, $P[a, b]$ with the norm of $C[a, b]$ is not complete. To see this, recall that by the Weierstrass approximation theorem,[13] every continuous function on a finite interval can be approximated with arbitrary precision by a polynomial. Thus, $P[a, b]$ is dense in $C[a, b]$. Because $P[a, b]$ contains a countable dense set, namely, the polynomials with rational coefficients, $C[a, b]$ is separable.

EXERCISES

1. The polynomials with rational coefficients are denumerable.
2. Prove $C[a, b]$ is complete.
3. Let c denote the space of convergent sequences. Thus if

$$x \in c, x = (x_1, x_2, \cdots, x_n, \cdots)$$

and $\lim_{n \to \infty} x_n$ exists. Prove that c is a complete metric space in the metric

$$d(x, y) = \sup\{|x - y_i| : i \in 1, 2, \cdots\}.$$

4. Prove the completeness of E_n.

B-4. Contraction Mapping Theorem

A map F of a metric space M into itself is a *contractor on M* if there is a number $q < 1$ such that for every pair of points x and y in M, $d(Fx, Fy) \leq qd(x, y)$. Let $F(Fx) = F^2x$, etc.

Theorem (fixed points of contractors)

Let M be a complete metric space and F a contractor on M. Choose x_0 arbitrarily in M. The sequence $\{F^nx_0\}$ converges to a point z such that $Fz = z$ and z is unique.

Proof

Let $F^nx_0 = x_n$. Choose $m \geq n$. Then

$$d(x_n, x_m) = d(F^nx_0, F^mx_0) \leq qd(F^{n-1}x_0, F^{m-1}x_0)$$

$$\leq q^nd(x_0, F^{m-n}x_0) = q^nd(x_0, x_{m-n}).$$

---

[13] Apostol, *op. cit.*, p. 481; Rudin, *op. cit.*, p. 147.

It is easy to see that

$$d(x_0, x_s) \le \sum_{i=1}^{s} d(x_{i-1}, x_i).$$

Thus,

$$q^n d(x_0, x_{m-n}) \le q^n \sum_{i=1}^{m-n} d(x_{i-1}, x_i).$$

We have shown that $d(x_{i-1}, x_i) \le q^{i-1} d(x_0, x_1)$ for $i \ge 1$. Thus,

$$d(x_n, x_m) \le q^n d(x_0, x_1) \sum_{i=1}^{m-n} q^{i-1}$$

But

$$\sum_{i=1}^{\infty} q^{i-1} = \frac{1}{1-q}.$$

Hence, $d(x_n, x_m) \le q^n d(x_0, x_1)/(1-q)$. Consequently, $\{x_n\}$ is a Cauchy sequence. Since M is complete, $\{x_n\}$ has a limit x in M. Since F is a contractor, it is continuous at x. Hence, $x = \lim_n \{x_{n+1}\} = \{\lim_n Fx_n\} = Fx$. That x is unique follows as in Section A-1.

Corollary

Let F be a map of a complete metric space M into itself. If, for some integer $n > 0$, F^n is a contractor on M, then F has a unique fixed point.

Proof

Since F^n is a contractor on M, it follows from the preceding theorem that F^n has a unique fixed point, say z. Hence, $Fz = FF^n z = F^n Fz$, and Fz is a fixed point of F^n. Hence, $Fz = z$.

Suppose z is not a unique fixed point of F. Then there exists $z_1 \ne z$ such that $Fz_1 = z_1$; then $F^n z_1 = z_1$. Hence, $z_1 = z$.

Remarks

In the preceding corollary, all that was needed to draw our conclusion was the uniqueness of the fixed point for F^n. Hence, we have also proved the more general theorem. Let S be any nonempty set of elements and F a map of S into itself. If, for some integer $n > 0$, F^n has a unique fixed point z, then F also has a unique fixed point, namely, z.

In Kolmogorov and Fomin,[14] the contraction mapping theorem is applied to the following problems:

---

[14] Kolmogorov and Fomin, *op. cit.*, p. 45.

(i)   Roots of equations;
(ii) Solutions of systems of linear equations;
(iii) Existence of the solution of differential equations and systems of differential equations;
(iv) Linear and nonlinear integral equations.

Fill in the details left out in the proof of the contraction mapping theorem by doing the following problems.

PROBLEMS

1. Prove that if $x_{n+1} = Fx_n$ and $z = Fz$, $d(x_n, z) \le q_n d(x_0, z)$.

2. If F is a contractor, F is continuous.

3. Show by induction that

$$d(x_0, x_n) \le \sum_{i=1}^{n} d(x_{i-1}, x_1).$$

Use the triangle inequality.

4. Use the triangle inequality to show that $\{x_n\} \to x$ and $\{y_n\} \to y$ imply $\{d(x_n, y_n)\} \to d(x, y)$.

B-5. Subcontraction Mapping Theorem (fixed points of subcontractors)

Let Q be a map of a metric space into itself such that

(i) $d(Qx, Qy) \le d(x, y)$;
(ii) if $x \ne Qx$, then $d(Qx, Q^2x) < d(x, Qx)$;
(iii) Q has a compact range.

Then for each x, the sequence $\{Q^n x\}$ converges to a fixed point of Q.

Proof

Appealing to (i), we may write

$$d(Q^n x, Q^{n+1}x) = d(QQ^{n-1}x, QQ^n x) \le d(Q^{n-1}x, Q^n x) \le \cdots (dx, Qx).$$

Thus, $\{d(Q^n x, Q^{n+1}x)\}$ is a nonincreasing sequence of real numbers bounded below by 0, and therefore has a limit. By (iii), the sequence $\{Q^n x\}$ belongs to a compact set. Thus, a subsequence $\{Q^{n_k}x\}$ has a limit point y. Since the sequence $\{d(Q^n x, Q^{n+1}x)\}$ converges, every subsequence—say $\{d(Q^{n_k}x, Q^{n_k+1}x)\}$ and $\{d(Q^{n_k+1}x, Q^{n_k+2}x)\}$—converges and has the same limit, namely,

$$d(y, Qy) = \lim_{k \to \infty} d(Q^{n_k}x, Q^{n_k+1}x) = \lim_{k \to \infty} d(Q^{n_k}x, QQ^{n_k}x).$$

We have used here the result of Problem 4 above and the continuity of Q. Thus,

$$d(y, Qy) = \lim_{k \to \infty} d(Q^{n_k+1}x, Q^{n_k+2}x) = \lim_{k \to \infty} d(QQ^{n_k}x, Q^2 Q^{n_k}x) = d(Qy, Q^2 y)$$

because Q^2 is also continuous. But this contradicts (ii) unless $y = Qy$. Given $\varepsilon > 0$, choose N such that $d(Q^N x, y) < \varepsilon$. Then

$$d(Q^{N+n}x, y) = d(Q^{N+n}x, Q^{N+n}y)$$
$$\leq d(Q^{N+n-1}x, Q^{N+n-1}y) \cdots \leq d(Q^N x, y) < \varepsilon,$$

by (i). Thus, $\{Q^n x\}$ converges to y.

SECTION C. MISCELLANY

C-1. Definitions

Our next project is to use fixed points of contractors and subcontractors to find roots of certain systems of nonlinear equations. To do this, we need a generalized mean-value theorem.

Let S be a subset of E_n such that for every pair of points x and y in S, the line segment joining x and y belongs to S. Everyone has agreed to call such sets "convex." Stated otherwise, if x and y are in S, then $\lambda x + \mu y \in S$ for λ, μ nonnegative and $\lambda + \mu = 1$. (*Examples:* In E_2, circle, triangle, square, line segment.) The convex hull of a set S is the intersection of all convex sets containing S.

A convex function f on E_n is a real-valued function defined on a convex subset S of E_n such that for x and y in S, $f(\lambda x + \mu y) \leq \lambda f(x) + \mu f(y)$, λ, $\mu \geq 0$, $\lambda + \mu = 1$.

Let $f \in C^1(S)$, where S is a convex subset of E_n with a nonempty interior. (If $f \in C^1(S)$, then the partial derivatives of f exist and are continuous on S.) Recall that the mean-value theorem of calculus states that

$$f(z) - f(y) = [\nabla f(\xi), z - y], \qquad (z, y \in S),$$

where ξ belongs to the open-line segment joining z and y,

$$\nabla f(\xi) = \left(\frac{\partial f(\xi)}{\partial x_1}, \frac{\partial f(\xi)}{\partial x_2}, \cdots, \frac{\partial f(\xi)}{\partial x_n} \right)$$

is an n-vector called the *gradient*, and

$$[\nabla f(\xi), z - y] = \sum_{i=1}^{n} \frac{\partial f(\xi)}{\partial x_i} (z_i - y_i).$$

We shall henceforth denote by $L(x, y)$ the open-line segment joining x and y. By the Schwarz inequality, $|[x, y]| \leq \|x\| \|y\|$, $|f(z) - f(y)| \leq M \|z - y\|$, where $M = \sup\{\|\nabla f(\xi)\| : \xi \in L(x, y)\}$. We shall generalize this inequality in Section C-3 to the case where f is *vector-valued* rather than real-valued.

C-2. Norms

We have seen that the distance function $d(x, y)$ on E_n, namely,

$$d^2(x, y) = \sum_{i=1}^{n} (x - y_i)^2$$

is a bona fide metric. The function $d(x, 0) = \sum [x_i^2]^{1/2} = \|x\| = [x, x]^{1/2}$ is called a *norm*. The norm measures the distance from a point to the origin. For example, $\|x - y\|$ is understood as follows: The vector x-y is the vector from the origin parallel to the directed-line segment from y to x. Thus, $\|x - y\|$ is the distance from the vector x-y to the origin. The norm, being a metric, satisfies

(i) $\|x\| = d(x, 0) = 0$ iff $x = 0$;
(ii) $\|x\| + \|y\| \geq \|x + y\|$ (triangle inequality).

To see that (ii) follows from the triangle inequality for metric spaces, we write

$$d(\xi, \eta) + d(\eta, \pi) \geq d(\xi, \pi) \Rightarrow \|\xi - \eta\| + \|\eta - \pi\| \geq \|\xi - \pi\|.$$

Put $x = \xi - \eta$, $y = \eta - \pi$. Then $\xi - \pi = x + y$. The function $x \to \|x\|$ above satisfies the condition

(iii) $\|\alpha x\| = |\alpha| \|x\|$ for all real numbers α.

These properties are then *abstracted*. Any function from E_n to the reals, satisfying (i), (ii), and (iii), is called a norm. Observe that $\|x\| \geq 0$ by (i), (ii), and (iii) because putting $x = -y$, $0 = \|x + y\| \leq \|x\| + \|y\| = 2\|x\|$.

Thus far we have considered norms on n-vectors. But norms on linear maps (or synonymously, linear operators, linear transformations) can also be defined, and their definitions are given in such a way that they have the same properties as the norm of a vector. Let A denote an $m \times n$ matrix, and x be an n-vector. A represents a linear map from E_n into E_m. Consider the inequality

$$\|Ax\| \leq M\|x\| \qquad \text{all } x \in E_n.$$

(The same symbol has been used here for possibly two different norms. Moreover, the norm on the left is defined on m-vectors and the right one on n-vectors.) We ask, can we calculate a number M for which this inequality is satisfied?

PROBLEM

Show that the inequality above is satisfied if M is taken to be

$$\left[\sum_{ij} \sum A_{ij}^2 \right]^{1/2}$$

where $A = (A_{ij})$. (*Hint:* Use Schwarz's inequality.)

We define $\|A\|$ to be the least number M for which the above inequality is satisfied. Observe that the number $\|A\|$ exists because the set of reals from which M can be chosen is bounded below by 0. Thus, $\|A\| = \inf\{B: \|Ax\| \le B\|x\|$ for all $x \in E_n\}$.

Lemma

$$\|A\| = \sup\left\{\frac{\|Ax\|}{\|x\|} : x \ne 0\right\} = \sup\{\|Ax\| : \|x\| = 1\}.$$

Proof

Let

$$M = \sup_{x \ne 0} \frac{\|Ax\|}{\|x\|}.$$

Thus, $\|Ax\|/\|x\| \le M$ for all $x \in E_n$, $x \ne 0$, and therefore $\|Ax\| \le M\|x\|$ for all $x \in E_n$. Hence, $M \ge \|A\|$. Since $\|Ax\| \le \|A\|\|x\|$ for all $x \in E_n$, if $x \ne 0$, $\|Ax\|/\|x\| \le \|A\|$, and

$$M = \sup_{x \ne 0} \frac{\|Ax\|}{\|x\|} \le \|A\|.$$

Hence, $M \le \|A\| \le M$, and consequently $M = \|A\|$.

PROBLEM

Show that if E_{mn} denotes the vector space of all $m \times n$ matrices, then $\|A\| = \sup\{\|Ax\| : \|x\| = 1\}$ is a norm on E_{mn}.

Let F be a map from E_n into itself such that the components of F are in C^1 on E_n. The *Jacobian* of the map F at the point z in E_n is the matrix J with components $\{\partial F_i(z)/\partial x_j\}$, $(1 \le i \le n)$, $(1 \le j \le n)$. The Jacobian will be denoted by $J(z)$. Thus, $J(z)x$ will denote the product of the matrix $J(z)$ with the n-vector x.

C-3. *Generalized Mean-Value Theorem*

Let S be a convex subset of E_n. Assume that F possesses a Jacobian at each point of S. Then $\|Fz - Fy\| \le \sup\{\|J(\xi)\| : \xi \in L(z, y)\}\|z - y\|$, where $\|x\|^2 = [x, x]$.

Proof

Let us apply the mean-value theorem to the real-valued function

$$\sum_{i=1}^{n} F_i(y)u_i,$$

where the u_i are the components of some unit vector u. We have

$$\sum_{i=1}^{n}(F_i(y) - F_i(z))u_i = [\nabla(\Sigma F_i(\xi))u_i, y - z],$$

where $\xi \in L(y, z)$. Observe that ξ depends also on u. Thus,

$$\left| \sum_{i=1}^{n}(F_i(y) - F_i(z))u_i \right| = \left| \sum_i u_i[\nabla F_i(\xi), y - z] \right|$$

$$\leq \|u\| \left[\sum_i [\nabla F_i(\xi), y - z]^2 \right]^{1/2}$$

$$= \|J(\xi)(y - z)\| \leq \|J(\xi)\| \|y - z\|$$

$$\leq \sup\{\|J(\xi)\|: \xi \in L(y, z)\} \|y - z\|.$$

Assume $F(y) \neq F(z)$ and choose

$$u = \frac{F(y) - F(z)}{\|F(y) - F(z)\|}.$$

Then

$$|\Sigma(F_i(y) - F_i(z))u_i| = \|F(y) - F(z)\|$$

and the theorem is proved.

C-4. Spectral Bounds

Let A be an $n \times n$ matrix with real components. We denote by A^* the transpose of A, that is, the matrix formed by interchanging rows and columns of A. We call A *symmetric* if $A^* = A$. Let A be symmetric and consider the quadratic function

$$x \rightarrow [x, Ax] = x^*Ax = \sum_{i=1}^{n} \sum_{j=1}^{n} x_i A_{ij} x_j.$$

The numbers $\mu = \min\{[x, Ax]: \|x\| = 1\}$ and $\lambda = \max\{[x, Ax]: \|x\| = 1\}$ are called *bounds for the spectrum* of A. Because the function[15] $[\cdot, A\cdot]$ is continuous and the set $\{x: \|x\| = 1\} = S$ is compact, there are points in S where

---

[15] By $[\cdot, A\cdot]$ is meant the function whose value at x is $[x, Ax]$.

$[\cdot, A\cdot]$ achieves its extrema μ and λ. If $\mu > 0$, A is called *positive definite*; if $\mu = 0$, A is *positive semidefinite*; if $\mu < 0$ and $\lambda > 0$, A is *indefinite*. Negative definite and negative semidefinite are similarly defined.

E X E R C I S E

Prove that the function $x \to [x, Ax]$ is continuous.

Theorem

Assume that A is symmetric. Then $\sup\{|[x, Ax]| : \|x\| = 1\} = \sup\{\|Ax\| : \|x\| = 1\} = \|A\|$.

Proof

By "Schwarzing," we get $|[x, Ax]| \le \|x\|\|Ax\| \le \|A\|$ for all x, $\|x\| \le 1$. Let $\sup|[x, Ax]| = \sup\{|\mu|, |\lambda|\} = \alpha$. Hence, $\alpha \le \|A\|$. Indeed, $\alpha = \|A\|$. To see this, we perform the following computation:

$$\|Ax\|^2 = \tfrac{1}{4}[4[Ax, Ax]] = \tfrac{1}{4}[2[Ax, Ax] + 2[A^2x, x]]$$

$$= \tfrac{1}{4}\{\gamma^2[Ax, x] + [Ax, Ax] + [A^2x, x] + \gamma^{-2}[A^2x, Ax]$$

$$- [\gamma^2[Ax, x] - [Ax, Ax] - [A^2x, x] + \gamma^{-2}[A^2x, Ax]]\}$$

$$= \tfrac{1}{4}[[A(\gamma x + \gamma^{-1}Ax), \gamma x + \gamma^{-1}Ax] - [A(\gamma x - \gamma^{-1}Ax), (\gamma x - \gamma^{-1}Ax)]$$

$$\le \tfrac{1}{4}(\alpha\|\gamma x + \gamma^{-1}Ax\|^2 + \alpha\|\gamma x - \gamma^{-1}Ax\|^2) \le \frac{\alpha}{4}[2(\|\gamma x\|^2 + \|\gamma^{-1}Ax\|^2)]$$

$$= \frac{\alpha}{2}[\gamma^2\|x\|^2 + \gamma^{-2}\|Ax\|^2],$$

where γ is any positive number. By calculus, when $\|Ax\| \ne 0$, the minimum of this last function will be achieved when $\gamma^2 = \|Ax\|/\|x\|$, whence $\|Ax\|^2 \le \alpha[\|Ax\|\|x\|]$. Hence, $\|Ax\| \le \alpha\|x\|$, showing that $\alpha \ge \|A\|$. Thus, $\alpha = \|A\|$.

Q.E.D.

We shall now prove this theorem another way by use of the Lagrange multipliers.[16]

Lemma 1

Assume that A is symmetric. The vectors of unit length which maximize or minimize $[\cdot, A\cdot]$ are eigenvectors of A.

---

[16] Apostol, *op. cit.*, p. 154.

Proof

Let f and ϕ be real-valued functions of the class $C^1(E_n)$. A necessary condition that a point x in $\{x \in E_n : \phi(x) = 0\}$ minimize or maximize the function f is that there exists a number δ such that

$$\nabla f(x) + \delta \nabla \phi(x) = 0.$$

We now set $f(x) = [x, Ax]$ and $\phi(x) = 1 - \|x\|^2$. Then

$$\frac{\partial}{\partial x_k} f(x) = \frac{\partial}{\partial x_k} \sum_{i=1}^n x_i \sum_{j=1}^n A_{ij}x_j = \sum_{j=1}^n A_{kj}x_j + \sum_{i=1}^n x_i A_{ik} = 2\sum_{j=1}^n A_{kj}x_j,$$

since A is symmetric. Thus, $\nabla f(x) = 2Ax$ and $\nabla \phi(x) = -2x$. Therefore, the above equations imply the equation $Ax = \delta x$, subject to $\|x\| = 1$. Recall now that a vector x satisfying this equation is an eigenvector corresponding to the eigenvalue δ. Since there exist unit vectors that are extremals for $[\cdot, A \cdot]$, it follows that there exists x and δ satisfying $Ax = \delta x$ and $\|x\| = 1$. Furthermore, x is such an extremal.

Lemma 2

Assume that A is symmetric. The number $\pm \delta$ is an eigenvalue of $Ax = \delta x$ if and only if δ^2 is an eigenvalue of $A^*Az = \delta^2 z$.

Proof

Assume $A^*Az - \delta^2 z = (AA - \delta^2 I)z = 0$ with $\delta^2 > 0$. Then

$$(A - \delta I)(A + \delta I)z = 0 = (A + \delta I)(A - \delta I)z.$$

There are thus four possibilities: $(A + \delta I)z = 0$, $(A - \delta I)z = 0$, $(A - \delta I)[(A + \delta I)z] = 0$, or $(A + \delta I)[(A - \delta I)z] = 0$. In any event, either $+\delta$ or $-\delta$ is an eigenvalue of A. Furthermore, $\delta^2 = 0$ implies $[z, A^*Az] = 0$, so that $Az = 0$. For the converse, if $Ax = \delta x$, then $A^*Ax = \delta A^*x = \delta Ax = \delta^2 x$.

Q.E.D.

We are now in a position to re-prove the theorem, namely, that

$$\alpha = \max\{|[x, Ax]| : \|x\| = 1\} = \max\{\|Ax\| : \|x\| = 1\}.$$

As before, by "Schwarzing," we see that $\alpha \leq \|A\|$. On the other hand, for some x, $\|A\|^2 = \|Ax\|^2 = [x, A^*Ax] \leq \max\{\mu^2, \lambda^2\} = \alpha^2$, whence $\|A\| \leq \alpha$.

Q.E.D.

The following lemma is useful for the theorem below.

Lemma

Let A be a symmetric matrix with spectral bounds μ and λ. Then $\|I - \gamma A\| = \max\{|1 - \gamma\mu|, |1 - \gamma\lambda|\}$, where I is the identity matrix.

Proof

If $\gamma \geq 0$ and $\|x\| = 1$, $\gamma\mu \leq [x, \gamma A x] \leq \gamma\lambda$. Thus,

$$1 - \gamma\mu \geq [x, Ix] - [x, \gamma A x] = [x, (I - \gamma A)x] \geq 1 - \gamma\lambda.$$

Here I is the identity matrix. If $\gamma < 0$, μ and λ are interchanged in the above inequality. Hence, $|[x, (I - \gamma A)x]| \leq \max\{|1 - \gamma\mu|, |1 - \gamma\lambda|\}$, proving the lemma.

EXERCISE

Assume γ, $\lambda \geq 0$. Show that if $f(\gamma) = \max\{|1 - \gamma\mu|, |1 - \gamma\lambda|\}$, then $2/(\mu + \lambda)$ minimizes f.

C-5. Minimization of Some Functions

The *Hessian* of f at $u \in E_n$ is the matrix with components

$$\frac{\partial^2 f(u)}{\partial x_i \, \partial x_j} \qquad (1 \leq i \leq n), \quad (1 \leq j \leq n).$$

Theorem

Assume that f is a function in C^2 on E_n such that the spectrum of the Hessian of $f(x)$ is bounded below by $\mu > 0$ and above by λ for all $x \in E_n$. Define the set $I_\delta = [\delta, (2/\lambda) - \delta]$ with $0 < \delta \leq 1/\lambda$. Define the sequence $x^{k+1} = x^k - \gamma \, \nabla f(x^k)$ with $\gamma \in I_\delta$ and x^0 arbitrary. Then $\{x^k\}$ converges to a point z such that $f(z) \leq f(x)$ for all $x \in E_n$.

Proof

We shall seek a point z such that $\nabla f(z) = 0$. Let $Gy = y - \gamma \, \nabla f(y)$, that is,

$$G_i(y) = y_i - \gamma \frac{\partial f(y)}{\partial x_i}.$$

Clearly, if $\gamma \neq 0$, y is a fixed point of G if and only if y is a root of $\nabla f(y) = 0$. The derivative of $G_i(y)$ with respect to x_j may be written

$$\frac{\partial G_i(y)}{\partial x_j} = \frac{\partial y_i}{\partial x_j} - \gamma \frac{\partial^2 f(y)}{\partial x_j \partial x_i}.$$

Observe that $\partial y_i/\partial x_j = 0$ when $i \neq j$ and $\partial y_i/\partial x_j = 1$ if $i = j$. In matrix notation, $J_G(y) = I - \gamma H(y)$, where $J_G(y)$ denotes the Jacobian matrix of G at y, I is the identity matrix, and $H(y)$ is the Hessian matrix of f at y. Thus, $\|J_G(y)\| = \|I - \gamma H(y)\|$. By the preceding lemma, $\|J_G(y)\| \leq \max\{|1 - \gamma\mu|, |1 - \gamma\lambda|\}$. Verify that for $\gamma \in I_\delta$ $|1 - \gamma\lambda| \leq 1 - \delta\lambda$ and $|1 - \gamma\mu| \leq 1 - \delta\mu$. Hence, $\|J_G(y)\| \leq q < 1$, where $q = \max\{1 - \delta\lambda, 1 - \delta\mu\} = 1 - \delta\mu$, for all y in E_n.

By the result of Section C-3, p. 20.

$$\|Gy - Gz\| \leq \sup\{\|J_G(\xi)\|: \xi \in L(y, z)\}\|y - z\| \leq q\|y - z\|.$$

Hence, G is a contractor, and thus, by the results of Section B-4, p. 15 $x^{n+1} = G(x_n)$ converges to a fixed point z with the property that $\nabla f(z) = 0$. For some $\xi \in L(x, z)$, we have[17]

$$f(x) - f(z) = [\nabla f(z), x - z] + \frac{1}{2}[x - z, H(\xi)(x - z)] \geq \frac{\mu}{2} \|x - z\|^2.$$

It follows that z is the unique minimizer of f. As in Section B-2, the convergence is at least as fast as a geometric progression with ratio q.

Remark

If a function f has a positive semidefinite Hessian, then f is convex. The proof may be found in Chapter 3, Section B-9, p. 122.

Corollary

Let F be a map from E_n to itself such that throughout E_n, J_F is continuous, symmetric, positive definite, and has its spectrum bounded above and below by λ and μ, respectively. Then F has a unique root z, namely, $F(z) = 0$. The iteration $x^{n+1} = x^n - \gamma F(x^n)$ converges to z at the rate of a geometric progression whenever $\gamma \in [\delta, (2/\lambda) - \delta], 0 < \delta \leq 1/\lambda$.

---

[17] This is Taylor's theorem. See, e.g., Apostol, *op. cit.*, p. 124.

Example

Let A be an $m \times n$ matrix of rank n. Thus, $m \geq n$. Let y be an m-vector and x an n-vector. Let $F(x) = \|Ax - y\|$, where $\|\cdot\|$ is the Euclidean norm on E_n. We seek to minimize F. We have

$$F^2(x) = \sum_{i=1}^{m} \left[\sum_{j=1}^{n} A_{ij}x_j - y_i \right]^2.$$

Clearly, it is enough to minimize F^2. We calculate that

$$\frac{\partial F^2}{\partial x_k} = 2 \sum_{i=1}^{m} \sum_{j=1}^{n} [A_{ij}x_j - y_i] A_{ik}$$

and

$$\frac{\partial^2 F^2}{\partial x_s \, \partial x_k} = 2 \sum_{i=1}^{m} A_{is} A_{ik} = (2A^*A)_{sk}.$$

Obviously, A^*A is symmetric, but it is also positive definite because

$$[x, A^*Ax] = x^*A^*Ax = (Ax)^*(Ax).$$

If $x \neq 0$, then $Ax \neq 0$ because A has rank n; consequently, $(Ax)^*(Ax) > 0$. To get an upper bound for λ, we calculate

$$\left[\sum_{ij} (A^*A)_{ij}^2 \right]^{1/2} = \lambda_0$$

(See first problem in Section C-2, p. 19.) Thus, if $\delta \leq \gamma \leq (2/\lambda_0) - \delta$, with $0 < \delta \leq 1/\lambda_0$, the sequence $x^{k+1} = x^k - \gamma \nabla(F^2(x^k))$, with arbitrary x^0, converges to a point z satisfying $\nabla F(z) = 0$. Since z is unique, z minimize F.

EXERCISE

In the above theorem, verify that $|1 - \gamma\lambda| \leq 1 - \delta\lambda$, $|1 - \gamma\mu| \leq 1 - \delta\mu$ and that $\{x: f(x) \leq f(x^0)\}$ is bounded. Prove the above Corollary. Give a geometric interpretation of minimizing $\|Ax - y\|$. Prove z is unique.

SECTION D. GRADIENT TECHNIQUES

D-1. Heuristic Remarks

The preceding theorem was proved as a special case of a very general theorem, and is easily improved. The hypotheses assumed are too strong to have them very often fulfilled. But we can do better, as can be conjectured, if we consider the following heuristics.

The iteration we have discussed is $x^{k+1} = x^k - \gamma \, \nabla f(x^k)$. Because $\nabla f(x)$ is the direction in which f increases most rapidly in small neighborhoods of x, we have, for γ sufficiently small and $\nabla f(x^k) \neq 0$, that $f(x^k - \gamma \, \nabla f(x^k)) < f(x^k)$. Alternatively,[18] we may choose γ to minimize f, and this too will effect a decrease in f, provided $\nabla f(x^k) \neq 0$. If $f(x^k)$ is perpetually decreasing, the only points x^k that arise during the process of the iteration belong to the set $S = \{x \in E_n : f(x) \leq f(x^0)\}$. This suggests that we require a bound on the spectrum of H on S only. Indeed, one finds that if $\gamma_k \in [\delta, (2/\lambda_0) - \delta]$, where $\|H(x)\| \leq \lambda_0$ for all $x \in S$; then, $\{f(x^k)\}$ is a decreasing sequence of numbers. If f is bounded below, then $f(x^k)$ converges downward to a limit and $\{\nabla f(x^k)\}$ converges to 0. The same phenomena occur if γ is chosen at each cycle by the method of steepest descent. Furthermore, there is no need that $H(x)$ be positive definite on S.

Lemma

Let f be defined on a neighborhood of a point z in E_n. Assume that ∇f is defined at z and $\nabla f(z) \neq 0$. Then, for some $z' \in E_n$, $f(z') < f(z)$.

Proof

Let $u(\gamma) = z - \gamma \, \nabla f(z)$, $g(\gamma) = f(u(\gamma))$, and calculate that

$$g'(\gamma) = \left[\nabla f(u(\gamma)), \frac{du}{d\gamma} \right] \quad \text{and} \quad g'(0) = -\|\nabla f(z)\|^2 < 0.$$

Because g is differentiable at 0, $(1/\gamma)(g(\gamma) - g(0)) - g'(0) < \varepsilon$ if $0 < \gamma < \delta$. Choose $\varepsilon = -(1/2)g'(0)$; assume $0 < \gamma < \delta$, and set $z' = z - \gamma \, \nabla f(z)$. Then

$$f(z') = g(\gamma) < \varepsilon \gamma + g(0) + \gamma g'(0) < g(0) = f(z).$$

D-2. Gradient Method

In what follows let f be defined on E_n and let S denote the level set of f at x^0; that is, $S = \{x \in E_n : f(x) \leq f(x^0)\}$. Let $H(x)$ denote the Hessian of f at x. The function f will be assumed to belong to the class $C^2(S)$.

By this is meant that the second partial derivatives of f exist and are continuous on S. Because S is not open, the derivatives will be assumed to exist on some open set containing S.

---

[18] This is usually called *steepest descent*.

Theorem

Assume that f is bounded below. Choose x^0 arbitrarily in E_n and assume that $f \in C^2$ on S. Assume that for some $\gamma_0 > 0$, $\|H(x)\| \le 1/\gamma_0$ for all $x \in S$. Choose δ and γ_k to satisfy $0 < \delta \le \gamma_0$ and $\delta \le \gamma_k \le 2\gamma_0 - \delta$, and set $x^{k+1} = x^k - \gamma_k \, \nabla f(x^k)$. Then:

  (i) The sequence $\{x^k\}$ belongs to S, $\{\nabla f(x^k)\}$ converges to 0, and $\{f(x^k)\}$ converges downward to a limit.

 (ii) Assume that S is bounded. Every cluster point z of $\{x^k\}$ satisfies $\nabla f(z) = 0$. If ∇f has a unique root on S, then $\{x^k\}$ converges to z and z minimizes f.

(iii) Assume there exists $\mu > 0$ such that $[y, H(x)y] \ge \mu\|y\|^2$ for all $y \in E_n$ and $x \in S$. Assume that S is convex. Then there is a unique point z minimizing f, and $\{x^k\}$ converges to z at the rate of a geometrical progression with constant $\le 1 - \mu\delta$.

(iv) Assume (iii) with $\mu \ge 0$ and S bounded. Let D denote the diameter of any sphere containing S. Then $f(x^k) - \|\nabla f(x^k)\| D \le f(z) \le f(x^k)$.

Proof

Let $x(\gamma) = x - \gamma \, \nabla f(x)$ and $\Delta(x, \gamma) = f(x) - f(x(\gamma))$. The point $\xi(\gamma)$ will be found in $L(x, x(\gamma))$.[†] Set $Q(\gamma) = [\nabla f(x), H(\xi(\gamma))\nabla f(x)]$. Assume $x \in S$ and $\nabla f(x) \neq 0$. By Taylor's theorem, we have, whenever $x(\gamma) \in S$, that

$$\Delta(x, \gamma) = \gamma\|\nabla f(x)\|^2\left[1 - \frac{\gamma}{2} Q(\gamma)\right].$$

We now show that $\delta \le \gamma \le 2\gamma_0 - \delta$ implies $x(\gamma) \in S$. Since $\nabla f(x) \neq 0$, we have by the lemma of Section D-1 that $\Delta(x, \gamma) > 0$ for γ sufficiently small. Let $\hat{\gamma}$ be the least positive γ for which $\Delta(x, \gamma) = 0$, if such exists. In this case $\Delta(x, \hat{\gamma}) = 0$ implies that

$$\left[1 - \frac{\hat{\gamma}}{2} Q(\hat{\gamma})\right] = 0,$$

which implies that $\hat{\gamma} = 2/Q(\hat{\gamma})$. Moreover, $x(\hat{\gamma}) \in S$, since $f(x) = f(x(\hat{\gamma}))$ and $x \in S$. Therefore, $\xi(\hat{\gamma}) \in S$ and $Q(\hat{\gamma}) \le 1/\gamma_0$, showing that $\hat{\gamma} \ge 2\gamma_0$. A fortiori $x(\gamma) \in S$ whenever $\delta \le \gamma \le 2\gamma_0 - \delta$. If $\hat{\gamma}$ does not exist, $x(\gamma) \in S$ for all positive γ. Thus, in either case, $\Delta(x, \gamma) \ge \gamma\|\nabla f(x)\|^2[1 - \gamma/2\gamma_0]$.

The function $\gamma[1 - (\gamma/2\gamma_0)]$ has a maximum at $\gamma = \gamma_0$ and a minimum value of $\delta[1 - (\delta/2\gamma_0)] > 0$, which is achieved at the values $\gamma = \delta$ and $\gamma = 2\gamma_0 - \delta$. Thus,

$$\Delta(x^k, \gamma_k) \ge \|\nabla f(x^k)\|\delta\left[1 - \frac{\delta}{2\gamma_0}\right].$$

---

†As before, $L(x,y)$ denotes the open line segment joining x and y.

Because f is bounded below and the values $f(x^k)$ are decreasing, the sequence $\{f(x^k)\}$ converges downward to a limit, say L. Consequently, $\{\Delta(x^k, \gamma_k)\}$ and $\{\nabla f(x^k)\}$ converge to 0.

To prove (ii), let $\{x^{k_i}\}$ be a subsequence converging to a cluster point z. Then, since $f \in C^2$ on S, ∇f is continuous at z; hence $\{\nabla f(x^{k_i})\}$ converges to $\nabla f(z) = 0$. If $\{x^k\}$ did not converge to z, there would be an infinite subsequence of $\{x^k\}$ which had a cluster point, say, $z' \neq z$. As above, $\nabla f(z') = 0$, contradicting that z alone satisfies $\nabla f(z) = 0$. Since S is compact and f continuous, there is a point, say z', minimizing f. Since a necessary condition that z' be a minimum is that $\nabla f(z') = 0$ (Exercise 1, below), it must be that $z = z'$.

To prove (iii), refer to Section C-5, replacing λ by $1/\gamma_0$ and E_n by S. The proof of Section C-5 shows that the map $G = I - \gamma \nabla f(\cdot)$ satisfies

$$\|Gy - Gz\| \leq q\|z - y\|$$

with $q < 1$ for $\gamma \in [\delta, 2\gamma_0 - \delta]$, while the proof of (i) above shows that G maps S into itself for $\gamma \in [\delta, 2\gamma_0 - \delta]$. Hence, G is a contractor. For the case when γ_k is not a fixed number in $[\delta, 2\gamma_0 - \delta]$, see Exercise 2, below.

To prove (iv), we observe by Taylor's theorem that

$$f(z) = f(x^k) + [\nabla f(x^k), z - x^k] + \tfrac{1}{2}[z - x^k, H(\xi)(z - x^k)], \qquad \xi \in L(x^k, z).$$

Thus, $f(x^k) \geq f(z) \geq f(x^k) - \|\nabla f(x^k)\| D$. These inequalities are useful to test when computations should terminate.

EXERCISES

1. If f has a gradient on an open set S, then a necessary condition that f have a maximum or minimum on S is that ∇f vanish somewhere on S.[19] In (ii) above, we claimed that a necessary condition that z' be a minimum was that $\nabla f(z') = 0$. But the set S is closed. Prove that $\nabla f(z') = 0$ is a necessary condition that z' be a minimum. Assume $x^0 \neq z'$.

2. Finish the proof of (iii) above. (*Hint:* First observe that by the mean-value theorem there is a matrix E tending to 0 with $x^n - z$ such that $\nabla f(x^n) = H(z)(x^n - z) + E(x^n - z)$.) (Warning! We cannot take $E = H(\xi) - H(z)$.) Then use the formula

$$\|x^{n+1} - z\| = \|x^n - z - \gamma_n \nabla f(x^n)\|.$$

3. Assume that every sequence $\{x^k\}$ satisfying $\{\|x^k\|\} \to \infty$ implies $\{f(x^k)\} \to \infty$. Prove that S is bounded.

4. Prove Remark (a) below.

---

[19] See Apostol, *op. cit.*, p. 132.

Remarks

(a) To prove (i) and (ii) of the preceding theorem, we need only assume that $f \in C^1$, and for each x and y in S,

$$\|\nabla f(x) - \nabla f(y)\| \le \frac{1}{2\gamma_0} \|x - y\|.$$

The proof is left as an exercise.

(b) For the case when S is bounded, suppose one knows the existence of some upper bound $1/\gamma_0$, but knows no numerical value for such. Then the following result can be used: for every $m \in (0, 1)$, there exists $\bar{\gamma} > 0$ such that

$$m = \frac{\Delta(x, \bar{\gamma})}{\bar{\gamma}\|\nabla f_k\|^2} \quad \text{and} \quad \Delta(x, \bar{\gamma}) \ge 2\|\nabla f_k\|^2 \gamma_0 m(1 - m).$$

Here and below $\nabla f_k = \nabla f(x^k)$.

Proof

Set $g(\gamma) = [1 - \gamma Q(\gamma)/2] = \Delta(x, \gamma)/\gamma\|\nabla f_k\|^2$. Since $g(0) = 1$, $g(\hat{\gamma}) = 0$, and g is continuous on $[0, \hat{\gamma}]$, g takes on all values in $[0, 1]$ on $[0, \hat{\gamma}]$. Take $\bar{\gamma}$ so that $g(\bar{\gamma}) = m$; then

$$\Delta(\bar{\gamma}) = \bar{\gamma}\|\nabla f_k\|^2 m \ge \bar{\gamma}\|\nabla f_k\|^2 \left[1 - \frac{\bar{\gamma}}{2\gamma_0}\right],$$

which implies that $\bar{\gamma} \ge 2\gamma_0(1 - m)$. Thus, $\Delta(x, \bar{\gamma}) \ge \|\nabla f_k\|^2 2\gamma_0 m(1 - m)$. Observe that the optimal a priori choice of m is $\frac{1}{2}$, but any m satisfying, say, $0 < \delta \le m \le 1 - \delta$ is satisfactory.

To employ the above results in practice, we lay down an interval, say $[\frac{1}{4}, \frac{3}{4}]$, $(\delta = \frac{1}{4})$, and seek by trial a value of γ for which $\frac{1}{4} \le g(\gamma) \le \frac{3}{4}$. This may be done, for example, by halving and doubling, linear interpolation or extrapolation, or both. Usually, a satisfactory value of γ will remain satisfactory for several iterations. Indeed, if m is sufficiently close to 1 so that γ happens to lie in the interval $(0, 2\gamma_0)$, this choice of γ is satisfactory at every step, provided $f \in C^2(S)$.

D-3. *Steepest Descent*

We have seen in the Remark (b) of Section D-2 that one could choose γ_k a posteriori, rather than a priori, from the interval $[\delta, 2\gamma_0 - \delta]$. This suggests that both the bound on the norm of the Hessian and the Lipschitz continuity of ∇f (Section D-2, Remark (a)) can be avoided. In what follows, we shall assume that f is in C^1 on S. In addition, we shall require that ∇f be uniformly continuous on S. Of course, if ∇f is continuous on S and S is compact, this will be automatically accomplished.

Let us again define $\Delta(x, \gamma) = f(x) - f(x - \gamma \nabla f(x))$, $(\gamma \geq 0)$;

$$g(x, \gamma) = \frac{\Delta(x, \gamma)}{\gamma \|\nabla f(x)\|^2} \quad \text{and} \quad S = \{x \in E_n : f(x) \leq f(x_0)\}.$$

Choose δ, $0 < \delta \leq \frac{1}{2}$.

Theorem 1

Assume that ∇f exists, is bounded, and is continuous on S. Take $\gamma_k < 1$ so that $\delta \leq g(x^k, \gamma_k) \leq 1 - \delta$ when $g(x^k, 1) < \delta$, or choose $\gamma_k = 1$ when $g(x^k, 1) \geq \delta$. Set $x^{k+1} = x^k - \gamma_k \nabla f(x^k)$. Then:

(i) If f is bounded below and ∇f is uniformly continuous on S, then $\nabla f(x^k)$ converges to 0, while $f(x^k)$ converges downward to a limit L.

(ii) If S is bounded, every cluster point of $\{x^k\}$, say z, satisfies $\nabla f(z) = 0$. If the cluster points of $\{x^k\}$ are finite in number, $\{x^k\}$ converges to one of them. If ∇f has a unique root, then it minimizes f.

(iii) If f and S are convex, then $L = \inf\{f(x) : x \in E_n\}$. If S is convex and bounded, every cluster point of $\{x^k\}$ minimizes f.

Proof

For small γ,

$$0 < \Delta(x, \gamma) = \gamma[\nabla f(\xi), \nabla f(x)] = \gamma\|\nabla f(x)\|^2 + \gamma[\nabla f(\xi) - \nabla f(x), \nabla f(x)],$$

where $\xi \in L(x, x - \gamma \nabla f(x))$. Thus,

$$g(x, \gamma) = 1 + \frac{[\nabla f(\xi) - \nabla f(x), \nabla f(x)]}{\|\nabla f(x)\|^2}.$$

Since ∇f is continuous at x and $\|\xi - x\| \leq \gamma\|\nabla f(x)\|$, $g(x, 0) = 1$. If $g(x, 1) < \delta$, then since $x - \gamma \nabla f(x) \in S$, if $g(x, \gamma) \geq 0$ and $\gamma \geq 0$, we infer that by continuity, g takes on all values between 1 and δ; there exists, therefore, a number $\gamma > 0$ so that $\delta \leq g(x, \gamma) \leq 1 - \delta$.

Set $\nabla f(x^k) = \nabla f^k$, assume that $\nabla f^k \neq 0$ and that $x^k \in S$. Then

$$\Delta(x^k, \gamma_k) = \gamma_k g(x^k, \gamma_k)\|\nabla f^k\|^2 \geq \gamma_k \delta\|\nabla f^k\|^2 > 0.$$

Thus, $f(x^{k+1}) < f(x^k)$ and $x^{k+1} \in S$. Assume that $\|\nabla f(x^k)\|^2 \nrightarrow 0$. There exists a subsequence $\{x^k\}$ and a number $\varepsilon > 0$ such that $\|\nabla f^k\|^2 \geq \varepsilon$. It follows that $\{\gamma_k\}$ is bounded away from 0; for if not, take a thinner subsequence $\{\gamma_k\}$, if necessary, such that $\{\gamma_k\} \to 0$. Since ∇f^k is bounded on S, $\|x^k - \xi^k\| \to 0$. By the uniform continuity of ∇f on S, it follows that $\{[\nabla f(\xi^k) - \nabla f(x^k),$

$\nabla f(x^k)]\} \to 0$, and therefore $g(x_k, \gamma^k) \to 1$, contradicting that $g(x_k, \gamma^k) \le 1 - \delta$ for all k. There exists, then, a number $q > 0$ such that $\gamma_k \ge q$. Hence, $\Delta(x^k, \gamma_k) \ge q\,\delta\varepsilon$, from which we contradict the hypothesis that f is bounded below. Thus, $\|\nabla f^k\|^2 \to 0$, showing (i).

As in Section D-2, if z is a cluster point of $\{x^k\}$, $\nabla f(z) = 0$. Thus, the number of roots of $f(z)$ is not less than the number of cluster points of $\{x^k\}$. Therefore, if ∇f has a unique root on S and S is bounded, $\{x^k\}$ converges to it.

If the roots of ∇f are finite in number, we may suppose the cluster points of $\{x^k\}$ to be finite also. Then, because $\{\|x^{k+1} - x^k\|\} \to 0$, it follows that the sequence $\{x^k\}$ actually converges to one of these cluster points. See Exercise 4 below. Observe that the same argument can be given if the roots of ∇f are all isolated points.

To prove (iii), we need the inequality $f(y) \ge f(x) + [\nabla f(x), y - x]$, which we shall show follows from the convexity of f. Thus, $[f(x + t(y - x)) - f(x)]/t \le f(y) - f(x)$ for all t in $[0, 1]$. But for some α in $(0, 1)$, we have that

$$[f(x + t(y - x)) - f(x)] = t[\nabla f(x + \alpha t(y - x), y - x].$$

Taking the limit as $t \to 0+$, we obtain the desired inequality. Suppose now that $L \ne \inf\{f(x^k)\}$. Choose $z \in S$ so that $f(z) < L$. Assume first that S is bounded. Then $0 > f(z) - f(x^k) \ge [\nabla f(x^k), z - x^k]$. Let u be a cluster point of $\{x^k\}$. Let $\{x^k\}$ be a subsequence converging to u; then $\{[\nabla f(x^k), z - x^k]\}$ converges to 0. Thus,

$$0 > \lim_{k \to \infty} [f(z) - f(x^k)] = f(z) - L \ge \lim_{k \to \infty} [\nabla f(x^k), z - x^k] = 0,$$

a contradiction. Hence, $L = \inf\{f(x): x \in E_n\}$, and every cluster point u of $\{x^k\}$ minimizes f.

Next we assume that S is unbounded. Again assume that we can choose z so that $f(z) < L$. Since

$$\|x^{k+1}\|^2 - \|x^k\|^2 = -2\gamma_k[x^k, \nabla f^k] + \gamma_k^2\|\nabla f^k\|^2,$$

it follows that if $\{\|x^k\|\}$ is unbounded, there is a subsequence $\{x^k\}$ such that $-[x^k, \nabla f^k] \ge -\gamma_k\|\nabla f^k\|^2/2$. Therefore, $[\nabla f^k, z - x^k] \ge [\nabla f^k, z] - \gamma_k\|\nabla f^k\|^2/2$. Thus, $0 > f(z) - f(x^k) \ge [\nabla f^k, z] - \gamma_k\|\nabla f^k\|^2/2$, from which we may derive a contradiction as before, since clearly $\{\gamma_k\|\nabla f^k\|^2\} \to 0$.

EXERCISES

Prove statements 1 to 4 below, assuming hypotheses of Theorem 1.

1. Let ϕ be a bounded map from S to E_n such that

$$[\nabla f(x), \phi(x)] \ge 0 \quad \text{and such that} \quad [\nabla f(x), \phi(x)] = 0$$

only if $\nabla f(x) = 0$. Set $g(x, \gamma) = \Delta(x, \gamma)/\gamma[\nabla f(x), \phi(x)]$. Assume that S is bounded. Take $\gamma_k < 1$ so that $\delta \le g(\gamma_k, x^k) \le 1 - \delta$ and set $x^{k+1} = x^k - \gamma_k \phi(x^k)$. Then every cluster point z of $\{x^k\}$ satisfies $\nabla f(z) = 0$.

    2. Show that if f is defined throughout E_n, the hypothesis that S is convex may be deleted in statement (iii) of the above theorem.

    3. Let the sequence $\{\phi_k\}$ be defined as follows: take the rows of the identity matrix cyclically, namely: $\delta_1, \delta_2, \cdots, \delta_n, \delta_1, \delta_2, \cdots, \delta_n, \delta_1, \cdots$. Set $\phi_1 = u_1 \delta_1, \cdots, \phi_{n+1} = u_{n+1} \delta_1, \cdots$, where $u_n = \mathrm{sgn}[\delta_n, \nabla f(x^n)]$, if $[\delta_n, \nabla f(x^n)] \ne 0$. Set $x^{n+1} = x^n - \gamma_n \phi_n$, where γ_n is chosen as in Theorem 1. If $[\delta_n, \nabla f(x^n)] = 0$, set $x^{n+1} = x^n$. Show that if S is bounded, every cluster point of $\{x^n\}$ is a root of ∇f.

    4. Complete the proof of the statement: if the roots of ∇f are finite in number, $\{x^k\}$ converges to one of them.

Corollary (fixed points of gradient mappings)

    Let Q be a "gradient" map such that Q is the gradient of a real-valued function ϕ, which is defined throughout E_n. Assume $\phi(x) \le q\|x\|^2/2$, $q < 1$. If Q is continuous, it has a fixed point.

Proof

    Define $f(x) = ([x, x]/2) - \phi(x)$. Thus,

$$f(x) \ge (\|x\|^2/2) - q\|x\|^2/2 = \|x\|^2(1 - q)/2 \ge 0.$$

So f is bounded below. Furthermore, $\nabla f(x) = x - Q(x)$, so that ∇f is continuous. We claim that for every x^0 in E_n, the set $\{x \in E_n : f(x) \le f(x^0)\}$ is bounded. Indeed $f(x) \le f(x^0)$ implies $\|x\|^2 \le 2(1 - q)^{-1} f(x^0)$. Clearly, z is a fixed point of Q if and only if it is a root of ∇f. Thus, by the above theorem we can construct z.

Remark 1

    It will be later shown, Chapter III, Section B-12, p. 132, that if the Jacobian $J_Q(x)$ is symmetric for all x, there exists a real-valued function ϕ such that Q is the gradient of ϕ.

    We now consider choosing the numbers γ_k by the method of steepest descent. We shall assume that S is bounded. There exists, therefore, γ_k, which maximizes $\Delta(x^k, \gamma) = f(x^k) - f(x^k - \gamma \nabla f(x^k))$.

Theorem 2 (steepest descent)

Assume that ∇f is continuous on the convex hull of S, and that S is bounded. Choose $\gamma_k > 0$ to minimize $f(x^k - \gamma \nabla f(x^k))$ and set

$$x^{k+1} = x^k - \gamma_k \nabla f(x^k).$$

Then $\{\nabla f(x^k)\} \to 0$, and every cluster point of $\{x^k\}$ is a root of ∇f.

Proof

Let z be a cluster point of x^k; clearly, $\{f(x^k)\}$ converges downward to $f(z)$. We shall show that $\nabla f(z) = 0$. For suppose not, and let $\gamma_0 > 0$ minimize $f(z - \gamma \nabla f(z))$. Thus,

$$f(z) = f(z - \gamma_0 \nabla f(z)) + \eta \qquad \text{for some} \quad \eta > 0.$$

For every x^k,

$$f(x^k - \gamma_0 \nabla f(x^k)) = f(z - \gamma_0 \nabla f(z) + (x^k - z) + \gamma_0 (\nabla f(z) - \nabla f)(x^k)))$$

$$= f(z - \gamma_0 \nabla f(z)) + [\nabla f(\xi^k), x^k - z + \gamma_0 (\nabla f(z) - \nabla f(x^k))],$$

with ξ^k in $L(z - \gamma_0 \nabla f(z), x^k - \gamma_0 \nabla f(x^k))$.

Let $\{x^k\}$ be a subsequence converging to z. Then $\{\nabla f(\xi^k)\}$ converges to $\nabla f(z - \gamma_0 \nabla f(z))$ and $x^k - z + \gamma_0 [\nabla f(z) - \nabla f(x^k)]$ converges to 0. Hence, for k sufficiently large, ξ^k belongs to the convex hull of S and,

$$f(x^k - \gamma_0 \nabla f(x^k)) \le f(z - \gamma_0 \nabla f(z)) + \frac{\eta}{2} = f(z) - \frac{\eta}{2}.$$

But

$$f(z) < f(x^k - \gamma_k \nabla f(x^k)) \le f(x^k - \gamma_0 \nabla f(x^k)) \le f(z) - \frac{\eta}{2},$$

a contradiction; hence, $\nabla f(z) = 0$.

The idea of steepest descent may be modified. For example, suppose one chooses the least value of γ for which $d\Delta(x^k, \gamma)/d\gamma$ vanishes.

Theorem 3

Assume that ∇f is continuous on the bounded set S and choose γ_k to be the smallest zero of $(d/d\gamma)(f(x^k - \gamma \nabla f(x^k))$ on the ray $\{x^k - \gamma \nabla f(x^k): \gamma \ge 0\}$. Then (ii) of Theorem 1 can be asserted.

Proof

As above in Theorem 1, set $\Delta(x, \gamma) = f(x) - f(x - \gamma \nabla f(x))$. Assume that $\nabla f(x) \neq 0$. Then

$$\frac{d\Delta(x, \gamma)}{d\gamma} = [\nabla f(x - \gamma \nabla f(x)), \nabla f(x)],$$

and

$$\gamma g(x, \gamma) = \frac{\Delta(x, \gamma)}{\|\nabla f(x)\|^2}.$$

Let $\hat{\gamma}$ be the least positive root of

$$\frac{d}{d\gamma}(\gamma g(x, \gamma)) = \frac{[\nabla f(x - \gamma \nabla f(x)), \nabla f(x)]}{\|\nabla f(x)\|^2}.$$

Thus,

$$\hat{\gamma} g(x, \hat{\gamma}) = \int_0^{\hat{\gamma}} \frac{[\nabla f(x) - \gamma \nabla f(x), \nabla f(x)]}{\|\nabla f(x)\|^2} \, d\gamma.$$

Assume that $\{\nabla f(x^k)\}$ does not converge to 0. For each $x = x^k$, consider the preceding integrand as a function of γ. The integrand is 1 at $\gamma = 0$ and is 0 at $\hat{\gamma} = \gamma_k$. We shall choose γ_0 so that the integrand $I(x, \gamma)$ is $\geq \frac{1}{2}$ whenever $\gamma \leq \gamma_0$. Take a subsequence $\{x^k\}$, which converges to z. This sequence with its limit point is a compact subset of S, so that ∇f is uniformly continuous on this compactum, which we call K. There exists, therefore, positive numbers q and r such that for all $x \in K, q < \|\nabla f(x)\| < r$. By the uniform continuity of ∇f on S, given $\varepsilon = q/2$, there exists δ such that

$$\|\nabla f(x - \gamma \nabla f(x)) - \nabla f(x)\| < \frac{q}{2} < \frac{\|\nabla f(x)\|}{2},$$

whenever $\gamma \|\nabla f(x)\| < \delta$, for all $x \in K$. Choose $\gamma_0 = \delta/r$. Then, if $\gamma \leq \gamma_0$, $\gamma \|\nabla f(x)\| < \delta$. Thus, if $\gamma \leq \gamma_0$ and $x \in K$,

$$\|\nabla f(x - \gamma \nabla f(x)) - \nabla f(x)\| < \|\nabla f(x)\|/2.$$

Thus,

$$\frac{1}{2} > \frac{1}{\|\nabla f(x)\|} \|\nabla f(x - \gamma \nabla f(x)) - \nabla f(x)\|$$

$$\geq \frac{1}{\|\nabla f(x)\|} \left\| \left[\nabla f(x - \gamma \nabla f(x)) - \nabla f(x), \frac{\nabla f(x)}{\|\nabla f(x)\|} \right] \right\|$$

$$= \left| \left[\frac{\nabla f(x - \gamma \nabla f(x))}{\|\nabla f(x)\|}, \frac{\nabla f(x)}{\|\nabla f(x)\|} \right] - 1 \right|.$$

which proves that if $x \in K$ and $\gamma \leq \gamma_0$, $I(x, \gamma) > \frac{1}{2}$. It follows, therefore, that

$$\gamma_0 < \hat{\gamma}, \; \hat{\gamma}g(\hat{\gamma}) > \int_0^{\gamma_0} \frac{1}{2} \, d\gamma = \frac{1}{2}\gamma_0 \quad \text{and therefore} \quad \Delta(x^k, \gamma_k) \geq \frac{q^2\gamma_0}{2},$$

contradicting that f is bounded below on K.

Remark 2

Assume that Q is a mapping that is the gradient of a function f on E_n, which is bounded below. Assume S is bounded and Q is continuous, and apply steepest descent. Observe that we have replaced the problem of finding a root of Q on E_n by a sequence of problems, each one requiring the location of a root of Q on a ray; these roots are the points x^k and $\{Q(x^k)\} \to Q(z) = 0$.

D-4. *Acceleration*

Suppose $f(x) = x_1^2 + \beta x_2^2$. The level sets of f, namely, the sets of the form $S = \{x: f(x) \leq K\}$, are ellipses if $\beta > 0$. If $\beta = 1$, S is a circle, and for some γ, $x^1 = x^0 - \gamma \nabla f(x^0) = 0$ minimizes f. If $\beta = 1/100$, we have, on the other hand, a skinny ellipse, and the gradient does not generally point toward 0. This example shows that the effectiveness of the gradient method can possibly be improved by a "change of scale" in the independent variables, more generally by such a change at each iterative step with a general linear transformation rather than by a simple change of scale. Let B be a nonsingular n-square matrix and introduce the variables $y = (y_1, \cdots, y_n)^*$ by means of the equation $x = By$. Set $f(x) = f(By) = g(y)$. Then $\nabla g(y) = B^* \nabla f(x)$. (See Exercise 1 below.) Descent in the variables $(y_1, \cdots, y_n)^*$ is made via an equation of the form $y(\gamma) = y - \gamma \nabla g(y)$. The corresponding variables $(x_1, \cdots, x_n)^*$ satisfy $B^{-1}x(\gamma) = B^{-1}x - \gamma B^* \nabla f(x)$, implying that $x(\gamma) = x - \gamma BB^* \nabla f(x)$. This suggests that we consider an iteration of the form $x^{k+1} = x^k - \gamma_k B_k B_k^* \nabla f^k$. In the next theorem below, we set $B_k B_k^* = H^{-1}$, where $H(x^k)$ is the Hessian of f at x^k, as before. For this to be possible, H must be positive definite.

EXERCISES

1. Show that $\nabla g(y) = B^* \nabla f(x)$. *Hint:*

$$\frac{\partial g}{\partial y_k} = \sum_{i=1}^n \frac{\partial f}{\partial x_i} \frac{\partial x_i}{\partial y_k}.$$

2. Assume that f is a quadratic function on E_n, and $BB^* = H^{-1}$. We have that the Hessian $H(x)$ is constant. Prove that $\{y: g(y) \leq C\}$ is a

sphere with center $B^{-1}z$. Here z minimizes f. *Hint:* Expand f at z and set $x = By$.

In the next theorem we argue as in Remark (b) of Section D-2 except that we take

$$g(x, \gamma) = \Delta(x, \gamma)/\gamma[\nabla f(x), H^{-1}(x)\nabla f(x)].$$

Here, as before,

$$\Delta(x, \gamma) = f(x) - f(x - \gamma H^{-1}(x)\nabla f(x)), \quad \nabla f^k = \nabla f(x^k)$$

and $H_k^{-1} = H^{-1}(x^k)$; the remaining nomenclature of the theorem of Section D-2 will also be assumed. Let $\hat{S}$ denote any convex set containing S.

Theorem

Assume that f is in C^2 on $\hat{S}$ and f is bounded below. Assume for $\xi \in$ and $x \in E_n$ that the Hessian matrix $H(\xi)$ satisfies $\mu\|x\|^2 < [x, H(\xi)x] \le \lambda\|x\|^2$ with $\mu > 0$. Set

$$x^{k+1} = x^k - \gamma_k H_k^{-1} \nabla f^k.$$

Here γ_k is chosen so that for $\delta < \frac{1}{2}$, $0 < \delta \le g(\gamma_k) \le 1 - \delta$, with $\gamma_k = 1$, if possible. Then:

(i)   For arbitrary x^0, the sequence $\{x^k\}$ converges to a point z, which minimizes f.

(ii) There exists a number N such that if $k > N$, then $\gamma_k = 1$.

(iii) The convergence of $\{x^k\}$ to z is *superlinear*; that is, it is faster than any geometric progression.

Proof

We show first that

$$\Delta(x, \gamma) \ge [\nabla f(x), H^{-1}(x)\nabla f(x)]\left[1 - \frac{\gamma\lambda^2}{2\mu^2}\right].$$

To prove this, we shall use the inequality $(\|x\|^2/\lambda) \le [x, H^{-1}(\xi)x] \le \|x\|^2/\mu$, which follows from our hypothesis. By Taylor's theorem,

$$\Delta(x^k, \gamma) = [\gamma \nabla f^k, H_k^{-1} \nabla f^k] - \frac{\gamma^2}{2} [H_k^{-1} \nabla f^k, H(\xi(\gamma))H_k^{-1} \nabla f^k]$$

$$\ge \gamma[\nabla f^k, H_k^{-1} \nabla f^k]\left[1 - \frac{\gamma}{2} \|H_k^{-1} \nabla f^k\|^2 \frac{\|H(\xi(\gamma))\|}{(\|\nabla f^k\|^2)/\lambda}\right]$$

$$\ge \gamma[\nabla f^k, H_k^{-1} \nabla f^k]\left[1 - \frac{\gamma\lambda^2}{2\mu^2}\right],$$

provided γ is sufficiently small, $\xi(\gamma) \in L(x^k, x^k - \gamma H^{-1}(x^k)\nabla f^k)$, and $\nabla f^k \neq 0$. Again by Taylor's theorem, for u and x in S,

$$\frac{(f(x) - f(u))}{\|x - u\|} - \left[\nabla f(u), \frac{(x - u)}{\|x - u\|}\right] \geq \mu \frac{\|x - u\|}{2}.$$

For $x \in S$, suppose $f(x) \geq B$ and $\|x - u\| \geq \varepsilon$. Then

$$\frac{f(x) - f(u)}{\|x - u\|} \leq \frac{f(x^0) - B}{\varepsilon}.$$

Hence, S is bounded. There exists, therefore, a least positive number $\hat{\gamma}$ such that $g(x^k, \hat{\gamma}) = 0$, showing that γ_k can be chosen as claimed. Thus, $g(x^k, \gamma_k) \geq [1 - \gamma_k \lambda^2/2\mu^2]$. Therefore,

$$\gamma_k \geq \frac{2(1 - g(x^k, \gamma_k))\mu^2}{\lambda^2} \geq \frac{2\delta u^2}{\lambda^2}$$

and

$$\Delta(x^k, \gamma_k) = \gamma_k g(x^k, \gamma_k)[\nabla f^k, H^{-1}(x^k)\nabla f^k] \geq \frac{2\|\nabla f^k\|^2 \delta^2 \mu^2}{\lambda^3}.$$

It follows, therefore, that $\{\nabla f(x^k)\}$ converges to 0. Let z be a cluster point for $\{x^k\}$; then $\nabla f(z) = 0$. Thus, $f(x) - f(z) \geq \mu\|x - z\|^2/2$ and z is a unique minimizer for f; moreover, $\{x^k\}$ converges to z.

To prove (ii), we observe that by setting

$$H(\xi(\gamma)) = H(x) + H(\xi(\gamma)) - H(x),$$

we get

$$g(x, \gamma) = 1 - \frac{\gamma}{2} - \frac{\gamma[H^{-1}(x)\nabla f(x), (H(\xi(\gamma)) - H(x))H^{-1}(x)\nabla f(x)]}{2[\nabla f(x), H^{-1}(x)\nabla f(x)]}.$$

Thus

$$|g(x, 1) - \tfrac{1}{2}| \leq \frac{\|H^{-1}(x)\nabla f(x)\|^2 \|H(\xi(1) - H(x)\|}{(2\|\nabla f(x)\|^2)/\lambda} \leq \frac{\|H(\xi(1) - H(x)\|\lambda}{2\mu^2}.$$

Since $\tfrac{1}{2}$ belongs to $(\delta, 1 - \delta)$ and $\{\|H(\xi(\gamma_k)) - H(x^k)\|\}$ converges to 0 by the uniform continuity of H on S, the assertion (ii) is now clear, or it will be after the exercise below is completed.

To prove (iii), assume $n > N$ and write

$$\nabla f^k = \nabla f(z) + (H_k + E_k)(x^k - z).$$

Here E_k is a matrix whose elements tend to 0. (Exercise 2, Section D-2, p. 29.) Thus

$$x^{k+1} - z = x^k - z - H_k^{-1}(H_k + E_k)(x^k - z) = -H_k^{-1}E_k(x^k - z),$$

whence $\|x^{k+1} - z\| \leq \mu^{-1}\|E_k\|\|x^k - z\|$, and the claim (iii) follows because $\{\|E_k\|\}$ converges to 0.

E X E R C I S E

Let $\alpha_k = \lambda\|J(\xi(\gamma_k)) - J(x_k)\|/2\mu^2$. Show that γ_k satisfies the inequality

$$\frac{1 - g(x_k, \gamma_k)}{\frac{1}{2} + \alpha_k} \leq \gamma_k \leq \frac{1 - g(x_k, \gamma_k)}{\frac{1}{2} - \alpha_k}$$

and thus eventually γ_k can be taken to be any point that satisfies

$$2\delta < \gamma_k \leq 2(1 - \delta).$$

It is now easy to see that the sequence we have described is eventually Newton's method applied to the function ∇f. We consider ourselves seeking roots of ∇f; equivalently, we are trying to solve simultaneous systems of nonlinear equations, each equation representing a component of the gradient. Or we may regard the equation $\nabla f(x) = 0$ as an operator equation on the operator $\nabla f: E_n \to E_n$. We shall take the point of view of Section A-2: We approximate each component of ∇f by a linear approximation and solve the resulting linear system. The linear approximation taken is the first two terms of the Taylor series. Thus, we write:

$$\nabla f(x^{k+1}) = 0 = \nabla f(x^k) + H(x^k)(x^{k+1} - x^k).$$

Solving,

$$x^{k+1} = x^k - H^{-1}(x^k)\nabla f(x^k).$$

This iteration is Newton's method applied to the operator equation $\nabla f(x) = 0$. Newton's method will be studied in general in Chapter III, Section C-4, for operators that are not necessarily the gradient of a function. The crux of the matter is that the Jacobian matrix of the system that represents the operator equation may not be symmetric. In the case of the gradient operator, the Jacobian is the Hessian, which is symmetric if $f \in C^2$.

We have seen that it is not necessary to move in a gradient direction to generate a minimizing sequence for f (Section D-3, Exercise 2).

E X E R C I S E

Show that the following are choices of effective vectors:
(a) $\phi = H^{-1}(x)\nabla f(x)$;
(b) $\phi = \|\nabla f(x)\|\delta_0$, where $\delta_0 = \max[\nabla f(x), \delta_i]$ and δ_i are the rows of the identity matrix and their negatives. ("Effective" here means generating sequences that converge to roots of ∇f.)

D-5. Semicontinuity

We now consider some mathematical concepts that will be needed in D-6.

Definitions

Let $\{x_k\}$ be a sequence of real numbers. When the number

$$\lim_{K \to \infty} (\inf\{x_k : k > K\})$$

exists, it is called "$\liminf\{x_k\}$." When "inf" is replaced by "sup" and the limit exists, it is called "$\limsup\{x_k\}$." Recall that a real-valued function f on a metric space M is continuous at a point $x \in M$, if and only if for every sequence $\{x_k\} \to x$ and for arbitrary $\varepsilon > 0$, there exists a natural number K such that $k > K$ implies $\varepsilon > f(x_k) - f(x) > -\varepsilon$. If we require only that the inequality $f(x_k) - f(x) > -\varepsilon$ holds, f is said to be *lower semicontinuous* (l.s.c.). If f satisfies $\varepsilon > f(x_k) - f(x)$, it is called *upper semicontinuous* (u.s.c.). Thus, f is lower semicontinuous at z if and only if for every sequence $\{x_k\} \to z$, $\liminf\{f(x_k)\} \geq f(z)$. Corresponding definitions can be made with ε and δ neighborhoods. If f is both l.s.c. and u.s.c., it is continuous.

Theorem

Let f be a real-valued function defined on a subset D of a metric space. Then f is l.s.c. on D if and only if the set $S_M = \{x \in D : f(x) \leq M\}$ is closed for all M.

Proof

Assume that S_M is closed for all M. Take $z \in D$ and assume that f is not l.s.c. at z. Then, for some number M and some sequence $\{x_k\} \to z$,

$$\liminf\{f(x_k)\} < M < f(z).$$

Thus, a subsequence of $\{x_k\}$ belongs to S_M, and since S_M is closed, $z \in S_M$. Contradiction. Assume, then, that f is l.s.c. and that S_M is not closed for some M. Then there exists a sequence $\{x_k\} \in S_M$ such that $\lim x_k = z \notin S_M$. Thus, $f(z) > M$. But since f is l.s.c. at z, $M \geq \liminf\{f(x_k)\} \geq f(z) > M$, a contradiction.

Theorem

An l.s.c. function defined on a compact subset D of a metric space achieves its infimum.

Proof

Let $\{x_k\}$ be a sequence in D such that $\{f(x_k)\} \to \inf\{f(x): x \in D\} = L$. Let z be a cluster point for $\{x_k\}$. Then $\liminf\{f(x_k)\} = L \geq f(z)$. Since $z \in D, f(z) = L$.

D-6. Roots of Systems of Equations

In Section C-5 the contraction mapping theorem was applied to minimize a function $f \in C^2(E_n)$ whose Hessian had a spectrum that was uniformly bounded above and below throughout E_n. It was shown that the contraction mapping theorem also provided for the construction of a root of a system of nonlinear equations, provided the system had a Jacobian that was symmetric and positive definite, and had a uniformly bounded spectrum. In Section D-3, no condition was imposed on the spectrum, but the symmetry of the Jacobian was still necessary.

The object of this section is to remove the hypothesis of symmetry of the Jacobian for root finding. We shall assume the Jacobian of the nonlinear system to be nonsingular on a given region. Our attack will be via a generalized gradient method, which will permit arbitrary directions, as mentioned in Section D-2, and which will be phrased to include Newton's method as a special case.

The motivation of our discussion is the following simple idea of Cauchy. Suppose that the system considered for solution is of the form

$$R_i(x) = 0, \qquad (1 \leq i \leq n),$$

where each R_i is a differentiable function on E_n. Consider the function

$$g(x) = \sum_{i=1}^{n} R_i^2(x) = \|R(x)\|^2.$$

Assume that $\nabla g(z) = 0$ and that the Jacobian J of R at z is nonsingular. Then, since

$$\sum_{i=1}^{n} R_i(z) \nabla R_i(z) = 0, \qquad (1 \leq i \leq n),$$

and since the set $\{\nabla R_i(z): 1 \leq i \leq n\}$ is linearly independent, it follows that $R_i(z) = 0$, $(1 \leq i \leq n)$.

We shall now prove two lemmas which will be needed in the sequel.

Lemma 1

Let A be an n-square matrix of rank n. Then

$$0 < \|A^{-1}\|^{-1} \leq \|Au\| \quad \text{for all } u, \qquad \|u\| = 1.$$

Proof

$$1 = \|u\| = \|A^{-1}(Au)\| \leq \|A^{-1}\| \|Au\|.$$

Moreover,

$$0 < \|A^{-1}\|^{-1}.$$

Q.E.D.

Lemma 2

Let R be a map from E_n into itself. Set $f(x) = \|R(x)\|$. Let z be a root of f and $\{y^k\}$ a sequence whose limit is z. Assume $\nabla f(y^k)$ and $f(y^k) \neq 0$. Let $T = \{y^k : k = 1, 2, 3, \cdots\}$. Assume J^* is defined and continuous[20] on $\bar{T}$, and $J^{*-1}(z)$ exists. Then there exists a positive number q such that $\|\nabla f(x)\| \geq q$ for all $x \in T$.

Proof

Since

$$f^2(x) = \sum_{i=1}^{n} R_i^2(x), \qquad f(x)\nabla f(x) = \sum_{i=1}^{n} R_i(x)\nabla R_i(x).$$

Thus,

$$\nabla f(x) = \frac{J^*(x)R(x)}{\|R(x)\|} \qquad \text{for all } x \in T.$$

Set $g(x) = \|\nabla f(x)\|$. Then g is continuous at $x \in T$. Set

$$g(z) = \inf\left\{\frac{\|J^*(z)R(x)\|}{\|R(x)\|} : x \in T\right\}$$

Thus defined, g is l.s.c. at z. To see this, calculate that

$$g(x) - g(z) = \left\| J^*(z)\frac{R(x)}{\|R(x)\|} + [J^*(x) - J^*(z)]\frac{R(x)}{\|R(x)\|} \right\| - g(z).$$

Choose a sphere $\delta(z)$ with center z and radius δ such that if $x \in \delta(z)$, $\|J^*(x) - J^*(z)\| < \varepsilon$. Then

$$g(x) - g(z) \geq \left\| \frac{J^*(z)R(x)}{\|R(x)\|} \right\| - \|J^*(x) - J^*(z)\|$$

$$- g(z) \geq -\|J^*(x) - J^*(z)\| > -\varepsilon,$$

showing that g is l.s.c. at z. Suppose now that $\|J^{*-1}(z)\| = B$. Then $g(z) \geq B^{-1}$ by Lemma 1. Thus, g is positive and l.s.c. on $T \cup \{z\} = T'$. Since g (see Section

---

[20] We mean by this that each component of J^* is continuous. Recall that $\bar{T}$ means the closure of T.

D-3) achieves its minimum on T', there exists a positive number q such that $g(x) \geq q$ for all $x \in T$. Thus, if $x \neq z$, $\|\nabla f(x)\| \geq q$.

E X E R C I S E

Prove that if the components of J are continuous at x, then J is continuous in norm at x; that is, given $\varepsilon > 0$, there exists a neighborhood $\delta(x)$ such that y in $\delta(x)$ implies

$$\|J(x) - J(y)\| = \sup\{\|(J(x) - J(y))u\| : \|u\| = 1\} < \varepsilon.$$

In the next theorem we follow the notation of the preceding lemma. In addition, let S denote the level set of f at x^0, and let ϕ denote a bounded map from S to E_n with the following properties:

$$[\nabla f(x), \phi(x)] = -f(x) \quad \text{and} \quad \|\phi(x)\| \leq \frac{f(x)}{\eta\|\nabla f(x)\|}$$

for some η, $0 < \eta < 1$. Set

$$\Delta(x, \gamma) = f^2(x) - f^2(x + \gamma\phi(x)) \quad \text{and} \quad g(x, \gamma) = \frac{f^2(x) - f^2(x + \gamma\phi(x))}{2\gamma f^2(x)},$$

when $f(x) \neq 0$.

Theorem

Let R be a map from E_n to E_n with $R_i \in C^2(S)$, $1 \leq i \leq n$. Assume that S is bounded. If $f(z) = 0$, assume that $J^{*-1}(z)$ exists. If $f(x) \neq 0$, assume that $\nabla f(x) \neq 0$. Given δ, $0 < \delta < \frac{1}{2}$, choose γ_k so that $\delta \leq g(x^k, \gamma_k) \leq 1 - \delta$. Set $x^{k+1} = x^k + \gamma_k\phi(x^k)$. If $f(x^k) = 0$, set $x^{k+1} = x^k$. Then:
(i)   Every cluster point z of $\{x_k\}$ is a root of R.
(ii) $f(x^k)$ converges downward to 0 at the rate of a geometric progression.
(iii) If J is nonsingular on S, every root of R is isolated, and $\{x^k\}$ converges at the rate of a geometric progression.

Proof

If $f(x^k) \neq 0$, $k = 0, 1, 2, \cdots$, it is always possible to choose $\phi^k = \phi(x^k)$. For example, take

$$\phi^k = \frac{-f(x^k)\nabla f^k}{\|\nabla f^k\|^2},$$

where $\phi^k = \phi(x^k)$ and $\nabla f^k = \nabla f(x^k)$.

Then

$$[\nabla f^k, \phi^k] = -f(x^k) \quad \text{and} \quad \|\phi^k\| = \frac{f(x^k)}{\|\nabla f^k\|}.$$

Assume that $f(x^k) \neq 0$, for $k = 1, 2, \cdots$, and extract a converging subsequence from $\{x^k\}$. Calculate that

$$\Delta(x^k, \gamma) = -\gamma[\nabla f^2(x^k), \phi^k] - \gamma[\nabla f^2(\xi^k) - \nabla f^2(x^k), \phi^k],$$

where $\xi^k \in L(x^k, x^k + \gamma\phi^k)$. Because $R_i \in C^2(S)$, $(1 \leq i \leq n)$, and S is bounded, R_i and ∇R_i, $(1 \leq i \leq n)$, satisfy uniform Lipschitz conditions on S. It follows that ∇f^2 is also Lipschitzian on S. Thus, for some M, $\|\nabla f^2(x) - \nabla f^2(y)\| \leq M\|x - y\|$ for all x and y in S. Hence, if $\xi^k \in S$, we find (observing that $\|\xi^k - x^k\| \leq \gamma\|\phi^k\|$) that

$$\Delta(x^k, \gamma_k) \geq -2\gamma_k f(x^k)[\nabla f(x^k), \phi^k]$$

$$- \gamma M \|\xi^k - x^k\| \|\phi^k\| \geq 2\gamma f^2(x^k)\left[1 - \frac{\gamma M}{2\eta^2 q^2}\right].$$

where $0 < q \leq \|\nabla f(x)\|$. The existence of q is guaranteed by Lemma 2, when we observe that if the limit of the subsequence $\{x^k\}$ is not a root of f, then clearly, $q > 0$. Let $\hat{\gamma}$ be the least $\gamma > 0$ such that $\Delta(x^k, \gamma) = 0$. We may write

$$\Delta(x, \gamma) = 2\gamma f^2(x'')\left[1 - \frac{[\nabla f^2(\xi^k) - \nabla f^2(x^k), \phi^k]}{f^2(x^k)}\right].$$

Since

$$g(x^k, \gamma) = \frac{\Delta(x^k, \gamma)}{2\gamma f^2(x^k)},$$

we can argue as in Remark (b) of Section D-2, p. 30. That is, we have $g(x^k, 0) = 1$ and $g(x^k, \hat{\gamma}) = 0$. Choose $m \in (\delta, 1 - \delta)$ such that $g(x^k, \gamma_k) = m$. Then

$$\Delta(x^k, \gamma_k) = 2\gamma_k m f^2(x^k) \geq 2\gamma_k f^2(x^k)\left[1 - \frac{\gamma_k M}{2\eta^2 q^2}\right],$$

so that

$$\gamma_k \geq \frac{2(1 - m)\eta^2 q^2}{M} \quad \text{and} \quad \Delta(x_k, \gamma_k) \geq \frac{4\delta(1 - \delta)\eta^2 q^2 f^2(x^k)}{M},$$

It follows that $\{f(x^k)\}$ converges to 0.

Since

$$f_k^2 - f_{k+1}^2 = 2\gamma_k f_k^2 g(x^k, \gamma_k), \qquad f_{k+1} = f_k(1 - 2\gamma_k g(x^k, \gamma_k))^{1/2}$$

Thus,

$$f_{k+1} = f_0 \prod_{i=0}^{k} (1 - 2\gamma_i g(x^i, \gamma_i))^{1/2}.$$

We calculate that

$$1 - 2\gamma_k g(x^k, \gamma_k) = \frac{f_{k+1}^2}{f_k^2} = 1 - \frac{\Delta(x^k, \gamma_k)}{f_k^2}$$

$$\leq 1 - \frac{4\delta(1 - \delta)\eta^2 q^2}{M}$$

$$= \alpha^2 < 1.$$

Thus, $f_{k+1} \leq f_0 \alpha^k$, and $\{f_k\}$ converges to 0 at the rate of a geometric progression.

Let z denote any cluster point of the sequence $\{x^k\}$. We shall prove, if the Jacobian of R is nonsingular on S, that $Q(x)$, the Hessian of f^2 at x, is positive definite for every x in some neighborhood of z. Let

$$A(x) = 2 \sum_{i=1}^{n} \frac{\partial^2 R_i(x)}{\partial x_j \, \partial x_k} R_i(x) \quad \text{and} \quad B(x) = 2 \sum_{i=1}^{n} \frac{\partial R_i(x)}{\partial x_j} \frac{\partial R_i(x)}{\partial x_k}.$$

Then $Q(x) = A(x) + B(x)$, and $[B(x)u, u] = 2\|J(x)u\|^2$. Moreover, because the Hessian of R_i, $(1 \leq i \leq n)$, is continuous on S, A is continuous and $A(z) = 0$. Let y minimize $\|J^{-1}(\cdot)\|^{-1}$ on S and set $\|J^{-1}(y)\|^{-2} = \mu$. Take a neighborhood $N(z)$ such that $x \in N(z)$ implies that $\|A(x)\| \leq \mu$. Then, for any unit vector u, we have

$$[Q(x)u, u] \geq -\mu + 2\|J(x)\|^2 \geq -\mu + 2\|J^{-1}(x)\|^{-2} \geq \mu > 0.$$

Since $\nabla f^2(z) = 0$, $f^2(x) = f^2(x) - f^2(z) \geq (\mu/2)\|x - z\|^2$ for $x \in N(z)$, it follows that the root z is isolated and $\{x^k\}$ converges at the rate of a geometric progression.

EXERCISE

1. Sketch a proof that $\|J^{-1}(\cdot)\|$ is continuous.
2. Show that ∇f^2 satisfies a Lipschitz condition on S.

Choice of Directions

We have already seen that the gradient of f furnishes a vector that can be made to satisfy the hypotheses of the above theorem if its length is suitably adjusted. There are also other simple choices. For example: Assume that at x^k, $J^{-1}(x^k)$ exists and that

$$\|J^{-1}(x^k)\| \le \frac{1}{\eta}\|\nabla f(x^k)\|;$$

then ϕ^k may be chosen by the *Newtonian dictum* $J(x^k)\phi^k = -R(x^k)$. Clearly,

$$\|\phi^k\| < \frac{f(x^k)}{\eta}\|\nabla f(x^k)\|$$

and

$$[\nabla f(x^k), \phi^k] = -\left[\frac{J^*(x^k)R(x^k)}{\|R(x^k)\|}, \quad J^{-1}(x^k)R(x^k)\right] = -f(x^k).$$

If J is nonsingular on S, there exist constants M and N so that $\|J^{-1}(x)\| \le M$ and $\|\nabla f(x)\| \le N$ for all $x \in S$. Suppose that η had been chosen so that $\eta \le 1/MN$. Then

$$\|\phi(x)\| \le Mf(x) \le \frac{f(x)}{\eta N} \le \frac{f(x)}{\eta\|\nabla f(x)\|}.$$

Thus, if J is nonsingular on S, ϕ can always be chosen by the Newtonian dictum.

EXERCISE

Let δ^i, $(1 \le i \le n)$, denote the rows of the identity matrix and choose $\eta \le 1/\sqrt{n}$. Define ϕ^k as follows:

$$\phi^k = \frac{-\delta_{i_0}\nabla f(x^k)}{|[\nabla f(x^k), \delta_{i_0}]|},$$

where i_0 is chosen so that

$$|[\nabla f(x^k), \delta_{i_0}]| \ge |[\nabla f(x^k), \delta_i]| \quad (1 \le i \le n).$$

Prove that ϕ^k satisfies the hypotheses of the theorem.

Corollary

Assume the hypotheses of the above theorem. Assume also that $J(\cdot)$ is nonsingular on S, and that ϕ^k is chosen to satisfy the Newtonian dictum, namely, $\phi^k = -J_k^{-1}R_k$. Choose $\gamma_k = 1$ if $g(x_k, 1) > 1 - \delta$. Otherwise, choose γ_k so that $\delta \le g(x^k, \gamma_k) \le 1 - \delta$, with $\gamma_k = 1$, if possible. Then:
(a)   There exists a number N such that if $k > N$, then $\gamma_k = 1$;
(b)   The sequence $\{x^k\}$ converges superlinearly to a root of R.

Proof

By the above discussion and by the theorem, the sequence $\{x^k\}$ converges to a root of R; call it z. Assume $f(x^k) > 0$, $k = 1, 2, 3, \cdots$. As in the above theorem, let $Q(x)$ denote the Hessian of f^2 at x and let $Q(x^n) = Q_n$. We first prove that

$$\lim_{n \to \infty} \frac{[\phi^n, Q_n \phi^n]}{2f_n^2} = \lim \theta_n = 1.$$

We have already shown that

$$\lim_{n \to \infty} [u, Q_n u] = 2 \lim_{n \to \infty} \|J_n u\|^2.$$

Since $\{x^n\}$ converges and ϕ is continuous,

$$\lim_{n \to \infty} [\phi^n, Q_n \phi^n] = 2 \lim_{n \to \infty} \|J_n \phi^n\|^2.$$

Thus,

$$\lim \theta_n = \lim \frac{\|J_n J_n^{-1} R_n\|^2}{f_n^2} = 1.$$

Calculate that

$$\Delta(x^n, \gamma) = f^2(x^n) - f^2(x^n + \gamma \phi^n)$$

$$= -2\gamma f_n [\nabla f^n, \phi^n] - \frac{\gamma^2}{2} [\phi^n, Q_n \phi^n] - \frac{\gamma^2}{2} [\phi^n, (Q(\xi^n) - Q_n)\phi^n],$$

where $\xi^n \in L(x^n, x^n + \gamma^n \phi^n)$. Hence, if γ is small enough so that $\xi^n \in S$,

$$\Delta(x^n, \gamma) = f_n^2 \left[2\gamma - \gamma^2 \theta_n - \frac{\gamma^2 \|Q(\xi^n) - Q_n\|}{2\eta^2 \|\nabla f^n\|^2} \right],$$

where, as before, $q > 0$ bounds $\|\nabla f(x)\|$ below and $\|\phi^n\| \le f_n/\eta q$. Since $g(x^n, \gamma^n) > 0$, $\xi^n \in S$. Let

$$\theta_n = 1 + \alpha_n, \quad \frac{\|Q(\xi^n) - Q_n\|}{2\eta^2 \|\nabla f^n\|^2} = \beta_n, \quad \text{and} \quad \alpha_n + \beta_n = \lambda_n.$$

Then $g(x^n, \gamma) = 1 - \gamma/2 - \gamma\lambda_n/2$. Since $\{\lambda_n\} \to 0$, for some N, $n \ge N$ implies $\delta < g(x^n, 1) < 1 - \delta$. Therefore, for some N, $n \ge N$ implies $\gamma_n = 1$. Thus, if $n = N$, $f_n^2 - f_{n+1}^2 = f_n^2[1 - \lambda_n]$, and $f_{n+1}^2 = \lambda_n f_n^2$ with $\{\lambda_n\} \to 0$. Observe that $0 \le \lambda_n < 1$. Therefore,

$$\|x_n - z\| \le \frac{2}{\mu} \left(\prod_{i=1}^{n-1} \sqrt{\lambda_i} \right) f_0 < \frac{2}{\mu} \sqrt{\lambda_n} \, f_0,$$

showing the superlinear convergence.

D-7. *Application to Linear Approximation*

We now consider as an application of the preceding theory some problems of linear approximation theory and smoothing of data. Let f be a bounded function defined on a set T. Suppose that we "sample" the values of f at m points of T, which we denote by $f(t_i)$, $(1 \leq i \leq m)$. Let $g_1, \cdots, g_n$ be n-bounded functions on T, which are also sampled at the points t_i $(1 \leq i \leq m)$. We consider approximating the sampled values of f by a linear combination of the sampled values of the functions $g_1, \cdots, g_n$. Let A denote the $m \times n$ matrix with components $A_{ij} = g_j(t_i)$, $(1 \leq i \leq m)$, $(1 \leq j \leq n)$, and b denote the m-vector with components $b_i = f(t_i)$, $(1 \leq i \leq m)$. Thus

$$\sum_{j=1}^{n} x_j g_i(t_i) - f(t_i) = \sum_{j=1}^{n} A_{ij} x_j - b_i.$$

Let $\| \cdot \|$ denote a norm on E_m. We define a *best approximation to f* by a linear combination of the functions g_j, $(1 \leq j \leq n)$, with respect to a $\| \cdot \|$, to be the function $\sum_{j=1}^{n} x_j g_j(t)$ if x minimizes $\|Ax - b\|$. We shall consider the 1_p norms defined as follows:

$$\|x\|_p = \left[\sum_{j=1}^{m} |x_j|^p \right]^{1/p}, \qquad p \geq 1.$$

Proof that $\| \cdot \|_p$ satisfies the triangle inequality is not trivial. The statement of the triangle inequality for this norm is called *Minkowski's inequality*.

Lemma (Hölder's inequality)

If $p > 1$, $(1/p) + (1/q) = 1$; then

$$\sum_{j=1}^{n} |x_j y_j| \leq \|x\|_p \|y\|_q.$$

Proof

We first prove that $a, b > 0$ implies that $ab \leq (a^p/p) + (b^q/q)$. To prove this, set

$$\phi(t) = \frac{t^p}{p} + \frac{t^{-q}}{q} \qquad \text{with } t > 0.$$

Then $\phi'(t) = t^{-1}(t^p - t^{-q})$. Since $p \geq 1$ and $q \geq 1$, $t < 1 \Rightarrow \phi'(t) < 0$. Thus, ϕ achieves a minimum at $t = 1$, and $\phi(1) = 1$. Set $t = a^{1/q}b^{-1/p}$. Then

$$\phi(t) = \frac{a^{p/q}b^{-1}}{p} + \frac{a^{-1}b^{q/p}}{q} \geq 1.$$

But $p/q = p - 1$ and $q/p = q - 1$; hence $(a^p/p) + (b^q/q) \geq ab$. Assume that $x, y \neq 0$ and let

$$u = \frac{x}{\|x\|_p} \quad \text{and} \quad v = \frac{y}{\|y\|_q}.$$

Since

$$\sum_{j=1}^{n} |x_j y_j| = \|x\|_p \|y\|_q \sum_{j=1}^{n} |u_j v_j|,$$

it will be sufficient to prove that $\sum_{j=1}^{n} |u_j v_j| \leq 1$. By the above,

$$u_j v_j \leq \frac{u_j^p}{p} + \frac{v_i^q}{q} \quad \text{if } u_j, v_j > 0.$$

Thus,

$$\sum |u_j v_j| \leq \frac{\sum |u_j|^p}{p} + \frac{\sum |v_j|^q}{q} = 1.$$

Q.E.D.

EXERCISE

$$\frac{a^p}{p} + \frac{b^q}{q} = ab \Leftrightarrow a^p = b^q \Leftrightarrow t = 1.$$

Show that this implies that equality holds in Hölder's inequality iff

$$|x_i|^p = \frac{\|x\|_p}{\|y\|_q} |y_i|^q, \quad j = 1, 2, \cdots, n \quad \text{(when } y \neq 0\text{)}.$$

Theorem (Minkowski inequality)

If $p \geq 1$, then

$$\|x + y\|_p \leq \|x\|_p + \|y\|_p.$$

Proof

$$[|a| + |b|]^p = [|a| + |b|]^{p-1}|a| + [|a| + |b|]^{p-1}|b|.$$

Thus,

$$\sum (|x_k| + |y_k|)^p = \sum (|x_k| + |y_k|)^{p-1}|x_k| + \sum (|x_k| + |y_k|)^{p-1}|y_k|.$$

Applying Hölder's inequality to the right side of this equality, we get

$$\sum (|x_k| + |y_k|)^p \le \left[\sum (|x_k| + |y_k|)^{(p-1)q}\right]^{1/q}\left[\sum |x_k|^p\right]^{1/p}$$
$$+ \left[\sum (|x_k| + |y_k|)^{(p-1)q}\right]^{1/q}\left[\sum |y_k|^p\right]^{1/p}.$$

Observing that $(p-1)q = p$ and dividing by $\left[\sum (|x_k| + |y_k|)^p\right]^{1/q}$, we get, since $1 - (1/q) = (1/p)$,

$$\|x + y\|_p \le \left[\sum (|x_k| + |y_k|)^p\right]^{1/p} \le \|x\|_p + \|y\|_p.$$

EXERCISES

Let

$$\|x\|_\infty = \max_{1 \le i \le m} |x_i|.$$

1. Show that $\|x\|_\infty \le \|x\|_p \le m^{1/p}\|x\|_\infty$, whence $\lim_{p \to \infty}\|x\|_p = \|x\|_\infty$.

2. Show that $\|x\|_r \le \|x\|_p$ if $p < r$. *Hint:* If $|x_j| = \|x\|_\infty$, write

$$\|x\|_p = |x_j|\left[1 + \sum_{i \ne j} \left|\frac{x_i}{x_j}\right|^p\right]^{1/p}.$$

3. Show that equality holds in Minkowski's inequality iff $x = cy$ for any $c \ge 0$. *Hint:* In the proof of the preceding theorem, examine carefully the application of Hölder's inequality.

Remark

For some constant C, $\|x\|_p \le C\|x\|_2$, $1 \le p \le \infty$.

We now return to the problem of minimizing $\|Ax - b\|_p$. We assume that the $m \times n$ matrix A has rank n.

EXERCISE

Show that no generality is lost by the assumption in the remark, provided $m \ge n$. *Hint:* All that matters is the column span of A.

Theorem

There exists a unique minimizer of $\|Ax - b\|_p$.

Proof

We prove first that there exists a minimizer of $\|Ax - b\|_p$. Consider the level set $L = \{x : \|Ax - b\|_p \le K\}$. This set is bounded. To prove this, it will be sufficient to show that for every sequence $\{x^k\}$ such that $\{\|x^k\|_p\} \to \infty$, $\{\|Ax^k - b\|_p\} \to \infty$. (See Section D-2, Exercise 3, p. 29.) Choose x^0 to minimize $\|Ax\|_p/\|x\|_p$ and set $\mu^0 = \|Ax^0\|_p/\|x^0\|_p$. Evidently $\mu_0 > 0$. Thus,

$$\|Ax - b\|_p \geq \|Ax\|_p - \|b\|_p \geq \mu_0\|x\|_p - \|b\|_p,$$

and therefore $\{\|x^k\|_p\} \to \infty$ implies that $\{\|Ax^k - b\|\} \to \infty$, showing that L is bounded. Clearly, if $\|Ax - b\|_p$ achieves a minimum, it will be achieved on L. Since $\|\cdot\|_p$ is a norm, the function $x \to \|Ax - b\|_p$ is continuous, and it achieves a minimum on L. We now prove that this minimum is unique.

By Exercise 3, on Minkowski's inequality, we know that $\|x + y\|_p = \|x\|_p + \|y\|_p$ holds only if $x = cy$. Suppose there are two solutions, x and y. Thus, $\|Ax - b\|_p = \|Ay - b\|_p = m$ and

$$\left\|A\frac{x+y}{2} - b\right\|_p = \left\|\frac{1}{2}(Ax - b) + \frac{1}{2}(Ay - b)\right\|_p$$

$$\leq \frac{1}{2}\|Ax - b\|_p + \frac{1}{2}\|Ay - b\|_p = m.$$

Hence, this last inequality must be an equality. It follows that $(Ax - b) = K(Ay - b)$. Clearly, $K = \pm 1$. If $K = 1$, $A(x - y) = 0$ and $x = y$. If $K = -1$, then

$$A\left(\frac{x+y}{2}\right) - b = 0 \quad \text{and} \quad m = 0.$$

There exists, therefore, again by the nonsingularity of A, a unique x such that $Ax - b = 0$. This contradiction shows that the minimizer is unique.

We shall assume with no loss of generality that the columns of A have unit length in $\|\cdot\|_2$.

Lemma

Let $f(x) = \|Ax - b\|_p^p$. Let $Q(x)$ denote the Hessian of f at x. Then

$$\|Q(x)\| \leq mnp(p - 1)[f(x)]^{1-(2/p)}.$$

Proof

Choose u^1 and u^2 so that $\|A\|_1 = \|Au^1\|$ and $\|A\|_2 = \|Au^2\|_2$. Then $\|Au^1\|_2 \leq \|Au^2\|_2 \leq \|Au^2\|_1 \leq \|Au^1\|_1$, since $\|x\|_2 < \|x\|_1$ by Exercise 1. Thus, $\|A\|_2 \leq \|A\|_1$. We calculate further that

$$\|A\|_1 = \|Au^1\|_1 = \max\left\{\sum_{i=1}^{m}\left|\sum_{j=1}^{n} A_{ij}u_j\right| : \|u\|_1 = 1\right\}$$

$$\leq \max\left\{\sum_{i=1}^{m}\max_j|A_{ij}|\sum_{j=1}^{n}|u_j| : \|u\|_1 = 1\right\}$$

$$= \sum_{i=1}^{m}\max_j|A_{ij}|.$$

Differentiating, we calculate that

$$Q_{jk}(x) = p(p-1) \sum_{i=1}^{m} |R_i(x)|^{p-2} A_{ij} A_{ik},$$

where

$$R_i(x) = \sum_{j=1}^{n} A_{ij} x_j - b_i.$$

Thus,

$$\|Q(x)\|_2 \le \|Q(x)\|_1 \le p(p-1) \sum_{j=1}^{n} \max_k \left| \sum_{i=1}^{m} |R_i(x)|^{p-2} A_{ij} A_{ik} \right|$$

$$\le p(p-1) \sum_{j=1}^{n} \max_k \max_i |A_{ij} A_{ik}| \sum_{i=1}^{m} |R_i(x)|^{p-2}$$

$$\le np(p-1) \|R(x)\|_{p-2}^{p-2}$$

$$\le mnp(p-1) [f(x)]^{1-(2/p)},$$

since

$$\sum_{j=1}^{n} \max_k \max_i |A_{ij} A_{ik}| \le n, \qquad f(x) = \|R(x)\|_p^p,$$

and by Exercise 1 above, $\|x\|_{p-2} \le (m^{1/(p-2)}) \|x\|_p$.

Thus, the algorithms of Section D-2 or Section D-3 may be applied to obtain converging sequence to the minimizer of f. For the algorithms of Section D-2, we also require that $p = 2$ or $p \ge 3$ in order for f to be in C^2. For the algorithm of Section D-3, it will suffice if $p > 1$.

We next show that under certain circumstances the algorithm of Section D-4 may also be applied.

Lemma

Let A' denote the matrix A augmented by the column b. Assume that every subset of $n+1$ rows of A' has rank $n+1$ and $m \ge 2n+1$. Then $Q(x)$ has rank n for all x.

Proof

Let A^i denote the ith row of A and let I denote a subset of $\{1, 2, \cdots, m\}$ containing $n+1$ or more integers. Suppose that $[A^i, x] = b^i$ for $i \in I$. Then $[A'^i, \bar{x}] = 0$ for $i \in I$, where $\bar{x} = (x, -1)$ and A'^i is the ith row of A'. This contradicts the linear independence of the rows of A'. There is, therefore, at least one element $i \in I$ such that $R_i(x) \neq 0$. Let I' be a subset of $n + 1$ integers of $\{1, 2, \cdots, m\} \sim \{i\}$. Again there is an integer $i' \in I'$ such that $R_{i'}(x) \neq 0$. It is now clear that R has $n + 1$ components that are nonzero. Since the rows of A corresponding to these components are of rank n, the matrix $Q(x)$ has rank n for all x. Q.E.D.

EXERCISES

1. Every subset of $n + 1$ rows of A in the lemma above has rank n.

2. Let $B = \{b_{ij} : 1 \leq i \leq m, 1 \leq j \leq n\}$ be an $m \times n$ matrix of rank n. Let d_i, $1 \leq i \leq m$ be nonzero numbers. If

$$C = \left\{ \sum_{i=1}^{m} d_i a_{ij} a_{ik} : 1 \leq i, j \leq m \right\},$$

then C has rank n.

By examining the formula for $Q(x)$, it is readily seen that $Q(x)$ is positive semidefinite. It follows, therefore, that $Q(x)$ is positive definite for all x and that the algorithm of Section D-4 can be applied if the hypotheses of the lemma are satisfied.

EXERCISES

1. Let P_{n-1} be a polynomial of degree $n - 1$, namely,

$$P_{n-1}(t) = x_0 + x_1 t + \cdots + x_n t^{n-1}.$$

Let $g(t) = P_{n-1}(t) + x_n e^t$. Show that $g(t)$ has at most n roots on $[a, b]$. *Hint:* Show that $g^{(n-1)}(t)$ has at most one root, $g^{(n-2)}$ at most two roots, and g at most n roots. (Here $g^{(n)}$ is the nth derivative of g.)

2. Let $[a, b]$ be sampled at m distinct points $t_1, t_2, \cdots, t_m$ in $[a, b]$. Consider the problem of minimizing

$$\sum_{i=1}^{m} \left| e^{t_i} - \sum_{j=0}^{n-1} x_j t_i^j \right|^p.$$

Show that the above lemma applies to this situation and that the algorithm of Section D-4 can be used.

3. In so-called uniform approximation, the norm $\| \cdot \|_\infty$ is used. That is, we seek a point x^∞ to minimize $\|Ax - b\|_\infty$. Assume that x^∞ exists and is unique. Let x^p minimize $\|Ax - b\|_p$. Prove that $\lim_{p \to \infty} x^p = x^\infty$. Let x^1 minimize $\|Ax - b\|_1$. Assume x^1 exists and is unique. Prove that $\lim_{p \to 1} x^p = x^1$. Note that of all the p norms, $\|x\|_\infty$ is least changed by small components of x and that $\|x\|_1$ is the most changed.

CHAPTER II

CONSTRAINTS

SECTION A. NONLINEAR PROGRAMMING

A-1. Constraints and Penalty Functions

In Chapter I, Section C-4 we considered the problem of minimizing the quadratic function $x \to [x, Ax]$ on the set $\{x: \phi(x) = \|x\| - 1 = 0\}$. To carry out this project, we used the method of Lagrange multipliers. The Lagrange-multiplier theorem requires that the function ϕ, whose level set defines the constraint, be differentiable. However, in many applications we wish to minimize or maximize on sets that cannot be represented as the level set of a differentiable function. We may also wish to find roots on such sets. Recall that the problem of determining whether a set S contains a root of a map R from S to E_n can be viewed as a problem of minimizing $\|R\|$ on S. If $\|R\|$ achieves a minimum on S and the minimum has the value 0, then R has a root in S; on the other hand, if the infimum of $\|R\|$ on S is positive, R does not have a root in S.

An example of a set that is not the level set of a differentiable function is a *polytope*. A polytope is the set of all points that satisfy a finite system of *linear inequalities* or simply inequalities. A finite system of linear inequalities may be written as follows:

$$[A^i, x] \le b^i \qquad (1 \le i \le m).$$

Here, for each i, $A^i \in E_n$ and b^i is real. To obtain some geometrical apprecia-
tion for such a system of linear inequalities, we consider a simple example in
the space E_2. Fix A in E_2. The level set $\{x \in E_2 : [A, x] = K\}$ is a line. (Recall
that in E_n the set $\{x \in E_n : [A, x] = k\}$, called a *hyperplane*, is $n - 1$ dimensional; that is, it contains a sphere of dimension at most $n - 1$.) We know from
calculus that the gradient of $[A, \cdot]$ at x is A, and that from x, A points in the
direction that $[A, \cdot]$ is increasing most rapidly. This is illustrated below.

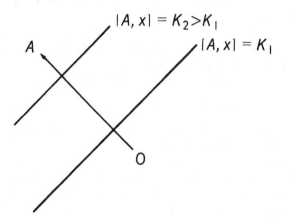

In these pictures we cannot, of course, represent the infinite extent of these
level sets. Below, the shaded region represents the points satisfying the

inequality $[A, x] \le K_2$. This set of points is called a *closed half-space*. Thus, the shaded region is the solution set of the inequality $[A, x] \le K_2$. The *solution of a system of linear inequalities* is that set which is the intersection of the half-spaces determined by the inequalities. If the half-spaces corresponding to a system of inequalities have a void intersection, the system has no solution.

EXERCISES

1. Write a system of three inequalities in E_2 whose solution set is a (nonempty) triangle.

2. Write a system of inequalities in E_3, which has no solution. (Such systems are said to be inconsistent.)

3. Let $f(x) = \max\{[A^i, x] - b^i : 1 \le i \le m\}$. Here, for each i, $A^i \in E_n$ and $b^i \in E_1$. By consideration of the level set $\{x : f(x) \le K\}$, characterize those points at which ∇f exists and does not exist.

4. Let $Q_i(x) = [A^i, x]$, $(1 \le i \le m)$. Consider the set

$$K = \{x : \alpha_i \le Q_i(x) \le \beta_i, 1 \le i \le m\}.$$

Assume K to be nonvoid. Show that K is bounded if and only if the set $\mathscr{A} = \{A^i : 1 \le i \le m\}$ spans n-space.

5. Let $S(M) = \{x \in E_n : [A^i, x] - b^i \le M, 1 \le i \le m\}$. Show that if $S(M)$ is bounded and nonempty for some M, it is bounded for all M.

6. Let $F_i(x) = 2[Q_i(x) - (\beta_i + \alpha_i)/2]/(\beta_i - \alpha_i)$. Show that the set K in Exercise 4 is equal to the set $\{x : -1 \le F_i(x) \le 1, 1 \le i \le m\}$. (Assume $\alpha_i \ne \beta_i$.)

7. Let P denote a bounded *polytope*, where by polytope we mean the intersection of a finite number of half-spaces. Show that there exists a family of *affine functions* F_i such that

$$P = \{x : -1 \le F_i(x) \le 1, 1 \le i \le m\}.$$

(An affine function on E_n is a linear function plus a constant, that is, $[A, \cdot] + b$, $A \in E_n$ and $b \in E_1$.)

8. If a hyperplane contains points x and y, it contains the line joining x and y. A hyperplane contains an $(n - 1)$ sphere, but not an n-sphere.

We now consider the problem of solving systems of linear inequalities by means of those techniques of minimizing functions which we have been discussing. Assume that we have a polytope P defined as in Exercise 6 and that we are trying to find a point in the polytope. Consider the expression

$$G_k(x) = \sum_{j=1}^{m} \frac{(F_j(x))^{2k}}{m}.$$

If $x \in P$, $|F_j(x)| \leq 1$, $1 \leq j \leq m$. If the inequalities are all strict, the terms $(F_j(x))^{2k}$ tend rapidly to 0 as k tends to ∞. If $|F_j(x)| \leq 1$ for all j, $1 \leq j \leq m$, then $0 \leq G_k(x) \leq 1$. On the other hand, if $x \notin P$, then for some j, $1 \leq j \leq m$, $|F_j(x)| > 1$ and $\lim_{k \to \infty} G_k(x) = \infty$. A function such as G_k is sometimes called a *penalty function*. It exacts a penalty—namely, large values—if it is evaluated outside P. On P it has small values. The above discussion suggests that if k is large, P is nonempty, and x^k minimizes G_k; then x^k is in P or close to P. We may state this formally in the following theorem and corollary.

Theorem

Let $P = \{x: -1 \leq F_j(x) \leq 1, 1 \leq j \leq m\}$, where each F_j is affine on E_n. Assume that P is bounded and nonvoid. Choose x^k so that

$$G_k(x^k) \leq G_k(x) = \sum_{j=1}^{m} \frac{(F_j(x))^{2k}}{m}$$

for all $x \in E_n$. Then every cluster point of $\{x^k\}$ lies in P.

Proof

To see that x^k can be so chosen, we argue as follows: Because P is bounded, if $\{\|x^i\|\} \to \infty$, $\{G_k(x^i)\} \to \infty$. Thus, for any fixed x^0 in E_n, the set $T = \{x: G_k(x) \leq G_k(x^0)\}$ is bounded. For the purpose of minimizing G_k, we may surely confine our attention to the set T. Since G_k is continuous, it has a minimum on T. Choose $x^1 \in P$. Then $|F_j(x^1)| \leq 1$, $(1 \leq j \leq m)$. Let

$$S_k = \left\{x: -1 - \frac{1}{k} \leq F_j(x) \leq 1 + \frac{1}{k}, 1 \leq j \leq m\right\}.$$

For fixed k and sufficiently large values of n, we have, if $x^n \notin S_k$,

$$1 \geq G_n(x^1) \geq G_n(x^n) > m^{-1}\left(1 + \frac{1}{k}\right)^{2n} > 1,$$

a contradiction. Hence, for large n, $x^n \in S_k$. Because P is bounded, we can conclude that S_k is bounded, arguing as in Exercise 4. Since S_k is also closed, any cluster point of $\{x^k\}$, say z, lies in S_k. This being true for all k, $z \in \cap S_k = P$. Observe that if P is a singleton, then $\{x^m\}$ converges.

Corollary

Choose x^m as in the above theorem. If P has a nonvoid interior, P^0, then for some N, $n \geq N$ implies that x^n lies in P^0.

Proof

Let $F(x) = (F_1(x), \cdots, F_m(x))$. Then,

$$\|F(x)\|_{2k}^{2k} = \sum (F_j(x))^{2k} = mG_k(x).$$

Hence, z minimizes G_k if and only if it minimizes $\|F(\cdot)\|_{2k}$. The function $\|F(\cdot)\|_\infty$ is continuous. By reasoning now familiar, the set $\{x: \|F(x)\|_\infty \leq K\}$ is closed and bounded for all K. Hence, $\|F(\cdot)\|_\infty$ achieves a minimum at, say, x^0. If x^* belongs to the interior of P, $\|F(x^*)\|_\infty < 1$. Thus, $\|F(x^0)\|_\infty \leq \|F(x^*)\|_\infty < 1$. Moreover, $\|F(x^0)\|_\infty \leq \|F(x^n)\|_\infty \leq \|F(x^n)\|_{2n} \leq \|F(x^0)\|_{2n}$. But $\lim_{n \to \infty} \|F(x^0)\|_{2n} = \|F(x^0)\|_\infty$, and hence for some N, $n \geq N$ implies that $\|F(x^0)\|_{2n} < 1$, which in turn implies that $\|F(x^n)\|_\infty < 1$, showing that x^n lies in P^0.

EXERCISE

Prove that if x satisfies $\|F(x)\|_\infty < 1$, then x is an interior point of P.

We have just seen that if P^0 is nonvoid, the above algorithm converges in a finite number of iterations; however, no upper bound is given for the total number of iterations. Furthermore, we have no way of checking whether P has an empty interior. For this reason we shall consider other methods of solving linear inequalities. However, the idea of the above theorem is useful for more general problems for which a variety of methods do not exist.

We now generalize in several directions. The finite family of affine functions $\{F_j: 1 \leq j \leq n\}$ will be replaced by a countably infinite family of real-valued nonlinear functions $\{F_j: j = 1, 2, 3, \cdots\}$ defined on a subset D of a metric space. The problem of finding a point satisfying $|F_j(x)| \leq 1$ for $j = 1, 2, 3 \cdots$ will be replaced by the problem of minimizing a lower semicontinuous (l.s.c.) function G on those points of D satisfying the inequalities $|F_j(x)| \leq 1, j = 1, 2, \cdots$. We form a sequence of functions by means of the expression

$$H_k(x) = G(x) + \left[\sum_{j=1}^{k} \frac{(F_j(x))^{2k}}{k^2} \right].$$

We shall assume outright that for some M, $k \geq M$ implies that H_k has a minimum on D, and that for some $k = N$, the set

$$S_k = \left\{ x \in D: -1 - \frac{1}{k} \leq F_j(x) \leq 1 + \frac{1}{k}, j = 1, 2, 3, \cdots \right\}$$

is compact and nonvoid. Set $S = \bigcap S_k$.

Theorem

Assume the above hypotheses. Assume that G is bounded below by $\mu > -\infty$. Let x^m minimize H_m. Then every cluster point of $\{x^m\}$ minimizes G on S.

Proof

Assume $x^0 \in S$. We have that $H_m(x^m) \le H_m(x^0) \le G(x^0) + 1$ for $m > M$. If $x_m \notin S_k$, then for some j, $|F_j(x^m)| > 1 + 1/k$. Thus,

$$H_m(x^m) > G(x^m) + \frac{1}{m^2}\left(1 + \frac{1}{k}\right)^{2m}.$$

Since $(1/m^2)(1 + (1/k))^{2m} > G(x^0) + 1 + |\mu|$ for all m sufficiently large, $H_m(x^m) > G(x^m) + G(x^0) + 1 + |\mu| \ge G(x^0) + 1$, a contradiction. Hence, $x^m \in S_k$ for large m. The existence of a cluster point follows from the compactness of S_k for $k \ge N$. Clearly, every cluster point of $\{x^m\}$ belongs to S. Let ξ be a cluster point for $\{x^m\}$ and η a point of S at which G attains its minimum over S. If $G(\xi) > G(\eta)$, set $2\varepsilon = G(\xi) - G(\eta)$. Since G is l.s.c. and $\{x^{m_i}\} \to \xi$, for some i sufficiently large, $G(x^{m_i}) - G(\xi) > -\varepsilon$. Choose $m \ge M$ so that $1/m < \varepsilon$ and $G(x^m) > G(\xi) - \varepsilon$. Since $\eta \in S$, $H_k(\eta) \le G(\eta) + 1/k$. Thus,

$$H_m(x^m) \ge G(x^m) > G(\xi) - \varepsilon = G(\eta) + \varepsilon \ge H_m(\eta) - \frac{1}{m} + \varepsilon > H_m(\eta),$$

contradicting that x^m minimizes H_m.

Remarks

(i) If G has a unique minimizer on S, then $\{x^k\}$ converges to it.
(ii)

$$G(x^m) - \frac{1}{m} \le G(\eta) \le G(x) \qquad \text{for all } x \in S.$$

Proof

To prove (ii) observe that if $z \in S$, then $H_k(z) \le G(z) + 1/k$. Hence,

$$G(x^k) - \frac{1}{k} \le H_k(x^k) - \frac{1}{k} \le H_k(z) - \frac{1}{k} \le G(z).$$

Set $z = \eta$. Q.E.D. Thus, given an estimate of $G(x^k)$, we may estimate a lower bound for $G(\eta)$. Observe that in the event that $x^m \in S$, an upper bound for $G(\eta)$ is also available.

Definition

A function f is said to be *strictly convex* on a convex subset S of E_n if, given x and y in S and $0 < t < 1$, $f(tx + (1 - t)y) < tf(x) + (1 - t)f(y)$.

EXERCISE

An l.s.c. strictly convex function f defined on a compact convex set achieves a unique minimum thereon. Prove this.

Remark

It will be shown in Chapter III, Section B-9 (lemma) that if the Hessian of f is positive definite on S, then f is strictly convex on S.

A-2. Extrema on Spheres and Supporting Hyperplanes for Convex Sets

Let S denote the unit sphere in E_n and δS the boundary of S. We now apply the techniques of Chapter I, Section D-3 to the problem of minimizing or finding "stationary" values of a differentiable function that is constrained to δS. In the event that ∇f is defined on S and does not vanish thereon, the constraint could be taken to be S rather than δS because in this instance the solution of our problem must be in δS. For this case the problem would be subsumed under Chapter III, Section B-11, where the case of minimizing on convex sets is discussed. (Clearly, S is convex, but δS is not.) However, in what follows, no assumption will be made about ∇f on the interior of S. The problem to be discussed is therefore not subsumed under Chapter III, Section B-11. As an application we consider a method for constructing a supporting hyperplane to a given boundary point of a convex set.

Assume that f is in C^1 on δS. Take x^0 arbitrarily in δS. Choose

$$3\mu \le \min\{\|\nabla f(x)\|^{-1} : x \in \delta S\}.$$

For $x \in \delta S$, set

$$x'(\gamma) = \frac{(x - \gamma \, \nabla f(x))}{\|x - \gamma \, \nabla f(x)\|},$$

$$\Delta(x, \gamma) = f(x) - f(x'(\gamma)).$$

and

$$g(x, \gamma) = \frac{-\Delta(x, \gamma)}{[\nabla f(x), x'(\gamma) - x]}.$$

Theorem

Choose δ, $0 < \delta \leq \mu$. Set

$$\theta_k = \min\{1, \tfrac{1}{3}|[\nabla f(x^k), x^k]|^{-1}\}.$$

Take $\gamma_k \leq \theta_k$ such that $\delta \leq g(x^k, \gamma_k) \leq 1 - \delta$ if $g(x^k, \theta_k) < \delta$ or $\gamma_k = \theta_k$ otherwise. Set

$$x^{k+1} = \frac{x^k - \gamma_k \nabla f(x^k)}{\|x^k - \gamma_k \nabla f(x^k)\|}.$$

Then:

(i) $\{\|\nabla f(x^k)\|^2 - [\nabla f(x^k), x^k]^2\}$ converges to 0, every cluster point z of the sequence $\{x^k\}$ satisfies $\nabla f(z) = \pm \|\nabla f(z)\| z$, $\{f(x^k)\}$ converges downward to a limit, and $\{x^{k+1} - x^k\}$ converges to 0.

(ii) Assume that for x in S, the roots of the equation

$$\nabla f(x) = \pm \|\nabla f(x)\| x$$

are finite in number. Then $\{x^k\}$ converges. If this equation has only one root in S, then $\lim\{x^k\}$ is a minimizer of f on S.

Proof

Assume that $\|\nabla f(x^k)\|^2 - [\nabla f(x^k), x^k]^2 \neq 0$. For simplicity, the index k will be dropped in some of what follows. Since $\|x\| = 1$, we find that

$$1 - \|x - \gamma \nabla f(x)\| = \gamma\left\{[\nabla f(x), x]\alpha(\gamma) + \frac{1 - \alpha(\gamma)}{\gamma}\right\}$$

where

$$\alpha(\gamma) = \left[1 + \frac{\gamma^2(\|\nabla f(x)\|^2 - [\nabla f(x), x]^2)}{(1 - \gamma[\nabla f(x), x])^2}\right]^{1/2}$$

and satisfies $\lim_{\gamma \to 0} \alpha(\gamma) = 1$ and $\lim(1 - \alpha(\gamma)/\gamma) = 0$; namely,

$$1 - \|x - \gamma \nabla f(x)\| = 1 - [(1 - \gamma[\nabla f(x), x])^2 + \gamma^2(\|\nabla f(x)\|^2 - [\nabla f(x), x]^2)]^{1/2}$$

$$= 1 - (1 - \gamma[\nabla f(x), x])\alpha(\gamma).$$

By the mean-value theorem,

$$\Delta(x, \gamma) = -[\nabla f(x), x'(\gamma) - x] - [\nabla f(\xi) - f(x), x'(\gamma) - x],$$

where ξ lies between x and $x'(\gamma)$. Thus,

$$g(x, \gamma) = 1 + \frac{[\nabla f(\xi) - f(x), f(x) - x\left([\nabla f(x), x]\alpha(\gamma) - \dfrac{1 - \alpha(\gamma)}{\gamma}\right)}{\|\nabla f(x)\|^2 - [\nabla f(x), x]\left([\nabla f(x), x]\alpha(\gamma) + \dfrac{1 - \alpha(\gamma)}{\gamma}\right)}$$

Because $\nabla f(x) \neq \|\nabla f(x)\| x$, $\|\nabla f(x)\|^2 > [\nabla f(x), x]^2$, and consequently, for γ sufficiently small, the denominator of the above fraction is positive. Consequently, since

$$\lim_{\gamma \to 0} \nabla f(\xi) = \nabla f(x), \quad \lim_{\gamma \to 0} g(x, \gamma) = 1.$$

Let $\sigma = [\nabla f(x), x]$ and $\varepsilon^2 = \|\nabla f(x)\|^2 - \sigma^2$. Since $\gamma > 0$ and $\gamma|\sigma| \leq \frac{1}{3}$,

$$1 < \alpha(\gamma) < 1 + \frac{\gamma^2 \varepsilon^2}{2(1 - \gamma|\sigma|)^2},$$

and observing that $1 - \alpha(\gamma) < 0$, we get

$$\|\nabla f\|^2 - \sigma\left(\sigma\alpha(\gamma) + \frac{1 - \alpha(\gamma)}{\gamma}\right) \geq \varepsilon^2 + \sigma^2(1 - \alpha(\gamma)) + |\sigma|\left(\frac{(1 - \alpha(\gamma))}{\gamma}\right)$$

$$\geq \varepsilon^2 - \frac{\sigma^2 \gamma^2 \varepsilon^2 + |\sigma|\gamma\varepsilon^2}{2(1 - \gamma|\sigma|)^2} \geq \frac{\varepsilon^2}{2}.$$

Thus, the denominator in the expression for g above is bounded away from 0, and therefore the function $\gamma \to g(x, \gamma)$ is continuous if $\gamma \leq \theta$. If $g(x, \theta) < \delta$, then this function takes on all values between δ and $1 - \delta$, so that γ_k is well defined. Since

$$g(x, \gamma) = 1 + \frac{[\nabla f(\xi) - \nabla f(x), x - x'(\gamma)]}{[\nabla f(x), x - x'(\gamma)]},$$

and we have shown that $\|x - \gamma \nabla f(x)\| [\nabla f(x), x - x'(\gamma)] \geq \varepsilon^2/2$, it follows that

$$[\nabla f(x), x - x'(\gamma)] \geq \frac{\varepsilon^2}{2(1 + \frac{1}{3}\mu)}.$$

Since $g(x, \gamma) \geq \delta$, $\Delta(x, \gamma) \geq (\delta\varepsilon^2/2)(1 + \frac{1}{3}\mu)$.

We now show the assumption $\{\|\nabla f(x^k)\|^2 - [\nabla f(x^k), x^k]^2\} \nrightarrow 0$ to be contradictory. By virtue of this assumption, we can choose a subsequence from $\{x^k\}$ such that

$$\{\|\nabla f(x^k)\|^2 - [\nabla f(x^k), x^k]^2\} = \{e_k^2\}$$

is bounded away from 0, say by $\bar{\varepsilon}$. It follows, therefore, that

$$\Delta(x^k, \gamma_k) \geq \frac{\delta \varepsilon^{-2}}{2(1 + \frac{1}{3}\mu)},$$

so that $\{f(x^k)\} \downarrow -\infty$. This contradiction establishes that

$$\{\|\nabla f(x^k)\|^2 - [\nabla f(x^k), x^k]^2\} \to 0.$$

It remains to prove that $\{x^{k+1} - x^k\} \to 0$. Clearly, we may assume $\{\nabla f(x^k)\} \to 0$. We calculate that

$$\frac{1}{2}\|x^{k+1} - x^k\|^2 = 1 - \frac{[x^k - \gamma_k \nabla f(x^k), x^k]}{\|x^k - \gamma_k \nabla f(x^k)\|}.$$

Set $\nabla f(x^k) = \mu_k x^k \|\nabla f(x^k)\| + \alpha^k$, where $\mu_k = \pm 1$, and recall that $\{\alpha^k\} \to 0$. Thus,

$$\frac{1}{2}\|x^{k+1} - x^k\|^2 = 1 - \frac{1 - \mu_k \gamma_k \|\nabla f(x^k)\| - \gamma_k[\alpha^k, x^k]}{\|x^k(1 - \mu_k \gamma_k \|\nabla f(x^k)\|) - \gamma_k \alpha^k\|}.$$

Since $|[\nabla f(x^k), x^k]| \geq \|\nabla f(x^k)\| - [\alpha^k, x^k]$ and $\gamma_k|[\nabla f(x^k), x^k]| \leq \frac{1}{3}$, for some $K > 0$, $k \geq K$ implies that $\gamma_k \|\nabla f(x^k)\| < \frac{1}{2}$. Hence, $k \geq K$ implies that

$$(1 - \mu_k \gamma_k \|\nabla f(x^k)\|) > \frac{1}{2}.$$

For large k, the numbers $\|x^k(1 - \mu_k \gamma_k \|\nabla f(x^k)\|) - \gamma_k \alpha^k\|$ are arbitrarily close to $1 - \mu_k \gamma_k \|\nabla f(x^k)\|$, showing that $\{x^{k+1} - x^k\} \to 0$.

Consider now the statements (ii) of the theorem. If $f(x) \pm \|\nabla f(x)\| x$ has m roots, then $\{x^k\}$ has no more than m cluster points; but if $m > 1$, a contradiction is easily reached because $\{x^{k-1} - x^k\} \to 0$. (See Chapter I, Section D-3, Exercise 4, p. 33.)

To conclude the proof of (ii) we observe: A necessary condition that z minimize f on δS is that $\nabla f(z) = \pm f(z)\|z\|$, since by the proof of (i), if this condition does not hold, then $f(z)$ can be decreased.

Example

To discuss the example we must first consider some technicalities. Let C denote a compact convex subset of E_n. Given an arbitrary vector $\xi \neq 0$, let $f(\xi) = \max\{[\xi, x]: x \in C\}$. The function f is called the *support function* for the set C. It is left as an exercise for the reader to verify that f is continuous, convex, and satisfies $f(m\xi) = mf(\xi)$ for all $m > 0$. The problem of finding a point $x \in C$, where $[\xi, x] = f(\xi)$, is called a problem in *linear programming*. Thus, linear programming is the problem of maximizing a linear function on a convex set. Techniques for solving such problems are considered in Section B.

Let x^* be a point satisfying $f(\xi) = [\xi, x^*]$. Then $[\xi, x^*] \geq [\xi, x]$ for all $x \in C$. The set $H = \{x \in E_n : [\xi, x^* - x] = 0\}$ is called a *hyperplane of support* (or *supporting hyperplane*) for C. This name is apropos because H and C meet at $x = x^*$ and the set C lies on one side of H. Thus, a supporting hyperplane is a hyperplane that meets the boundary of a convex set and contains the set in one of its half-spaces.

We call C "strictly convex" if the boundary of C, δC, contains no line segment.

Lemma

If C is a strictly convex set, then every hyperplane H of support for C meets δC in only one point.

Proof

If $x \neq y$ and $x, y \in H \cap C$, then $[\xi, x] = [\xi, y] = f(\xi)$. If $0 \leq t \leq 1$, then $[\xi, tx + (1 - t)y] = f(\xi)$. We know that both H (see Section A-1, Exercise 8, p. 56) and C are convex. Hence, $tx + (1 - t)y \in H \cap C$. But H meets C only in δC. Hence, $tx + (1 - t)y \in \delta C$, contradicting the strict convexity of C.

Let C be a compact strictly convex set. Then, for every $\xi \neq 0$, there exists a unique point $x(\xi)$ such that $[\xi, x(\xi)] = f(\xi)$. The mapping $x: \xi \to x(\xi)$ will be called the *support mapping* for the convex set C. Since $x(\xi) = x(\lambda \xi)$ for $\lambda > 0$, the natural domain of x is the set $\{\xi \in E_n : \|\xi\| = 1\}$. Thus, x maps the boundary of the unit sphere into the boundary of C.

EXERCISES

1. Prove the properties of the support function mentioned above.
2. Interpret all the above notions geometrically in E_2.
3. Prove the support mapping for a strictly convex set is onto. You may assume the following theorem: If C is a convex set with interior, then at every point $x \in \delta C$ there exists a hyperplane of support for C. (See Chapter III, Section C-1, Exercise 2, p. 137.)

Lemma

The support mapping for a compact strictly convex set is continuous.

Proof

Assume $\{\xi^i\} \to \xi$. By definition,

$$[x(\xi), \xi] \geq [x(\xi^i), \xi] \quad \text{and} \quad [x(\xi^i), \xi^i] \geq [x(\xi), \xi^i].$$

Thus,

$$[x(\xi), \xi] + [x(\xi^i), \xi^i - \xi] \geq [x(\xi^i), \xi^i] \geq [x(\xi), \xi^i].$$

Hence, $\lim\{[x(\xi^i), \xi^i]\} = [x(\xi), \xi]$. Since the function $x \to [x, \xi]$ achieves a unique maximum on C at $x(\xi)$, every cluster point of $\{x(\xi^i)\}$ equals $x(\xi)$, showing that x is continuous at ξ.

Lemma

The support mapping for a strictly convex compact set C is the gradient of the support function for C.

Proof

From calculus[1] we know that a function f has a gradient at a point ξ if, given $\varepsilon > 0$, there exists $\delta > 0$ and a linear function $h \to [\alpha, h]$ such that $|f(\xi + h) - f(\xi) - [\alpha, h]| < \varepsilon\|h\|$ whenever $\|h\| < \delta$. When an n-vector α exists with this property, it is called the *gradient of f* at ξ and is written $\nabla f(\xi)$. Let ϕ denote the support function for the convex set C. We must show that $|\phi(\xi + h) - \phi(\xi) - [x(\xi), h]| < \varepsilon\|h\|$. We calculate that

$$[\phi(\xi + h) - \phi(\xi)] = [x(\xi + h), \xi + h] - [x(\xi), \xi]$$

$$= [x(\xi + h) - x(\xi), \xi] + [x(\xi), h] + [x(\xi + h) - x(\xi), h]$$

Thus,

$$|\phi(\xi + h) - \phi(\xi) - [x(\xi), h]| \leq |[x(\xi + h) - x(\xi), \xi]| + |[x(\xi + h) - x(\xi), h]|$$

$$\leq 2|[x(\xi + h) - x(\xi), h]|.$$

To see that this last inequality is valid, observe that by the definition of the supporting hyperplane,

$$[\xi + h, x(\xi) - x(\xi + h)] \leq 0 \quad \text{and} \quad [-\xi, x(\xi) - x(\xi + h)] \leq 0.$$

Since the sum of these inequalities is less than or equal to either inequality,

$$[h, x(\xi) - x(\xi + h)] \leq [\xi, x(\xi + h) - x(\xi)] \leq 0,$$

---

[1] See Apostol, *op. cit.*, pp. 107, 111.

so that

$$|[h, x(\xi) - x(\xi + h)] \leq |[\xi, x(\xi + h) - x(\xi)]|.$$

To conclude, we observe that

$$2|[h, x(\xi) - x(\xi + h)]| \leq 2\|h\| \, \|x(\xi) - x(\xi + h)\|,$$

and by the continuity of x we can choose δ so that $\|h\| < \delta$ implies $\|x(\xi + h) - x(\xi)\| < \varepsilon/2$. This completes the proof.

The above two lemmas assert that ϕ has a continuous gradient $\nabla\phi$ and $\nabla\phi(\xi) = x(\xi)$. We now formulate more precisely the problem stated in the beginning of this section.

PROBLEM

Assume that for each ξ, $x(\xi)$ can be found. Obtain ξ^0, $\|\xi^0\| = 1$ such that $x(\xi^0) = x^0$; or, stated otherwise, determine the supporting hyperplane to C at x^0, with C strictly convex and compact.

EXERCISE

Let H be the hyperplane $\{y \in E_n: [\xi, y - \hat{x}] = 0\}$, $\|\xi\| = 1$, and x^0 a point of E_n. Show that the distance from x^0 to H is $|[\xi, \hat{x} - x^0]|$.

Let x^0 be a point on the boundary of C and set $\hat{x} = x(\xi)$. Then the above hyperplane H is a supporting hyperplane for C at $x(\xi)$. Set $\phi(\xi) = [\xi, x(\xi) - x^0]$. Stationary values of ϕ are values for which

$$\nabla\phi(\xi) = x(\xi) - x^0 = \pm\xi\|x(\xi) - x^0\|.$$

Since $[\xi, x(\xi) - x^0] \geq 0$, the negative sign cannot occur. Because $\phi(\xi)$ is the distance from x^0 to the hyperplane of support for C at $x(\xi)$, $\phi(\xi) \leq \|x(\xi) - x^0\|$. Let

$$S^0 = \left\{ \xi \in S: \frac{x(\xi) - x^0}{\|x(\xi) - x^0\|} = \xi \right\}.$$

We now prove that the set S^0 is compact. By the Lemma below, the set S^0 is isolated from the roots of the equation $x(\xi) - x^0 = 0$. Hence if $\xi \in S^0$, the mapping

$$\xi \rightarrow \left[\frac{x(\xi) - x^0}{\|x(\xi) - x^0\|}, \xi \right]$$

is a continuous map from S^0 to a single point, which is a closed set. It follows that S^0 is a closed subset of a compact set.

Lemma

The roots of the equation $x(\xi) = x^0$ are isolated from the set S^0.

Proof

The mapping $\xi \to x(\xi)$ is a continuous onto mapping from S to the boundary of C. (See Exercise 3 above.) Hence there exists $\hat{\xi} \in S$ such that $x(\hat{\xi}) = x^0$. Clearly $\phi(\hat{\xi}) = 0$. Suppose that the root $\hat{\xi}$ was not isolated from S^0. There exists then a sequence $\{\xi^k\}$, $\xi^k \in S^0$, such that $\{\xi^k\} \to \hat{\xi}$. But

$$\left[\xi^k, \frac{x(\xi^k) - x^0}{\|x(\xi^k) - x^0\|} \right] = 1,$$

while

$$\left[\hat{\xi}, \frac{x(\xi^k) - x^0}{\|x(\xi^k) - x^0\|} \right] < 0,$$

which is impossible for large k.

Theorem

Let $\mu = \min \{\phi(\xi) : \xi \in S^0\}$. Let ξ^0 be given satisfying $\phi(\xi^0) < \mu$. Every cluster point of the sequence generated by the algorithm above commencing at ξ^0 is a root of the equation $x(\xi) = x^0$. If C is uniquely supported at x^0, this sequence converges.

E X E R C I S E S

1. Draw examples in the plane of convex sets C so that S^0 has no points, one point, two points, and infinitely many points.

2. Assume a priori that a given component of $\hat{\xi}$ is nonzero. Show that the methods of Chapter I, Section D-3, will afford global convergence in the minimization of ϕ. *Hint*: Set the nonzero component of ξ equal to 1.

Remark

In the subject of optimal control theory (a sample problem is given in Chapter III, Section E-1), the solution of the linear programming problem of finding the value of $x(\xi)$ is given by maximizing a certain integral.

SECTION B. POLYHEDRAL CONVEX PROGRAMMING

Our object in this section is to sample some mathematical programming techniques that may be used to minimize convex functions which are not everywhere differentiable. In the same spirit, the convex sets that prescribe the domain of minimization may also have corners. We digress temporarily to examine some technical preliminaries.

B-1. On Homogeneous Linear Inequalities

Definition

The *convex hull* of a subset S of E_n, denoted by $H(S)$, is the set of all *convex combinations* of S, that is, the set of all finite linear combinations $\Sigma_i \lambda_i x^i$ where $x^i \in S$, $\lambda_i \geq 0$ and $\Sigma_i \lambda_i = 1$.

Remark

The set of all convex combinations of S is identical with the intersection of all convex sets containing S (Chapter III, Section A-1, Exercise 3, p. 94.)

Lemma (Carathéodory)

Let S be a subset of E_n. Every point in $H(S)$ may be expressed as a convex combination out of S, with at most $n + 1$ points.

Proof

Assume that $x \in H(S)$. Then for some integer k and $x^i \in S$,

$$x = \sum_{i=1}^{k} \lambda_i x^i \qquad \text{with } \Sigma \lambda_i = 1 \quad \text{and} \quad \lambda_i > 0.$$

Assume that the minimal such k is greater than $n + 1$. There exists, then, numbers $\alpha_1, \cdots, \alpha_k$, $\Sigma |\alpha_i| \neq 0$, such that $0 = \Sigma \alpha_i x^i$ and $\Sigma \alpha_i = 0$. This follows because the set $\{(x^i, 1) : 1 \leq i \leq k\}$ is linearly dependent. Because $\Sigma \alpha_i = 0$ and $\Sigma \lambda_i = 1$, the k-tuple $(\alpha_1, \cdots, \alpha_k) \neq m(\lambda_1, \cdots, \lambda_k)$ for any number m. Thus,

$$x = \sum_i (\lambda_i + t\alpha_i) x^i \qquad \text{for all } t.$$

For $|t|$ sufficiently small, $\lambda_i + t\alpha_i > 0$, $(1 \leq i \leq k)$. There exists, therefore, a number t_0, least in magnitude, such that for some index i_0, $\lambda_{i_0} + t_0 \alpha_{i_0} = 0$. If $i \neq i_0$, $\lambda_i + t_0 \alpha_i \geq 0$; moreover, for some index i, $\lambda_i + t_0 \alpha_i > 0$. We have therefore exhibited x as a convex combination out of S with less than k points. This contradiction establishes the lemma.

Lemma

The convex hull of a compact subset of E_n is compact.

Proof

Let A denote the compactum and A_n the $(n + 1)$-fold Cartesian product of A. By Caratheodory's lemma, each $x \in H(A)$ can be written as a sum:

$$\sum_{i=1}^{n+1} \lambda_i(x) A^i(x) \quad \text{with } \lambda_i(x) \geq 0 \quad \text{and} \quad \Sigma \lambda_i(x) = 1.$$

Let

$$Q = \left\{ (t_1, \cdots, t_{n+1}), \quad t_i \geq 0, \quad \text{and} \quad \sum_{i=1}^{n+1} t_i = 1 \right\}.$$

Q is compact. Thus, if $\{x^k\} \in H(A)$,

$$x^k = \sum_{i=1}^{n+1} \lambda^i(x^k) A^i(x^k)$$

with $(\lambda^1(x^k), \cdots, \lambda^{n+1}(x^k)) \in Q$ and $(A^1(x^k), \cdots, A^{n+1}(x^k)) \in A_n$.

Thus, we may extract a subsequence of $\{A^1(x^k), \cdots, A^{n+1}(x^k)\}$ converging to, say, $\{A^1, \cdots, A^{n+1}\}$, and a thinner subsequence of $\{\lambda_1(x^k), \cdots, \lambda_{n+1}(x^k)\}$ converging to, say, $\{\lambda_1, \cdots, \lambda_{n+1}\}$. Clearly,

$$\{x^k\} \to x = \sum_{i=1}^{n+1} \lambda_i A^i;$$

moreover, $A^i \in A, \lambda_i \geq 0, 1 \leq i \leq n + 1$, and

$$\sum_{i=1}^{n+1} \lambda_i = 1.$$

Lemma

Let S be a closed convex subset of E_n and let p be a point of E_n not in C. There exists a hyperplane separating p and C.

Proof

We show first that the distance from p to C is achieved. Take x^0 arbitrarily in S. If z minimizes $\|x - p\|$, then clearly $\|z - p\| \leq \|x^0 - p\|$. Thus, we concern ourselves with the set $\{x \in C : \|x - p\| \leq \|x^0 - p\|\} = K$. K is a closed subset of a sphere and hence is compact.

There exists, therefore, z in C nearest to p because the function $x \to \|x - p\|$ is continuous on K. We claim $[p - z, x - z] \leq 0$ for all $x \in C$. Take $x \in C$. By the convexity of C, $tx + (1 - t)z \in C, 0 \leq t \leq 1$. Thus,

$$\|tx + (1 - t)z - p\| - \|z - p\| \geq 0,$$

since z is nearest p; furthermore,

$$\|t(x - z) + z - p\|^2 \geq \|z - p\|^2 \Rightarrow t^2\|x - z\|^2 + 2t[x - z, z - p] \geq 0.$$

If $[x - z, z - p] < 0$, this inequality will be violated for small t. Hence, $[p - z, x - z] \leq 0$. Thus, if $x \in C, [p - z, x - z] \leq 0$; and if $x = p, [p - z, x - z] = \|p - z\|^2 > 0$. This shows that the hyperplane $\{x: [p - z, x - z] = 0\}$ separates p and C. Moreover, observe that this hyperplane supports C at z.

Lemma

Let $\mathscr{A}$ be a compact subset of E_n, and A a point of $\mathscr{A}$. The system of linear inequalities

$$[A, x] < 0 \qquad \text{all } A \in \mathscr{A} \quad (I)$$

is inconsistent iff $0 \in H(\mathscr{A})$.

Proof

Assume $0 \in H(\mathscr{A})$ and that (I) is consistent with solution x. Then

$$0 = \sum_{i=1}^{n+1} \lambda_i A^i \qquad \text{with } \lambda_i \geq 0 \ \Sigma\lambda_i = 1 \quad \text{and} \quad A^i \in \mathscr{A}.$$

Thus, $[\Sigma\lambda_i A^i, x] = 0 = \Sigma\lambda_i[A^i, x] < 0$. For the converse, assume $0 \notin H(\mathscr{A})$. By the separation lemma, p. 69, since $H(\mathscr{A})$ is closed, we find that $[A - z, -z] \leq 0$ for all $A \in H(\mathscr{A})$. Thus, $[A, -z] \leq -[z - z] < 0$ for all $A \in \mathscr{A}$, showing that $-z$ satisfies (I).   Q.E.D.

EXERCISE

Consider the following system of inequalities:

$$\left.\begin{array}{ll} [A, x] < 0 & \text{for all } A \in \mathscr{A}^0 \\ [A, x] \leq 0 & \text{for all } A \in \mathscr{A}^1 \end{array}\right\}(I)$$

Prove:
   (a) If $0 \notin H(\mathscr{A}^1)$, then $0 \in H(\mathscr{A}^0 \cup \mathscr{A}^1)$ implies that (I) is inconsistent.
   (b) Assume $H(\mathscr{A}^0 \cup \mathscr{A}^1)$ is closed. Then (I) inconsistent implies that $0 \in H(\mathscr{A}^0 \cup \mathscr{A}^1)$.

B-2. Polyhedral Convex Programming

Let S be a subset of E_n. If every subset of n-points of S is of rank n, then S is said to satisfy the *Haar condition*.

Lemma

Let $S = \{A^1, \cdots, A^{n+1}\}$ be an n-dimensional subset of E_n. Assume that $0 \in H(S)$. The matrix whose columns are $(A_j^1, A_j^2, \cdots, A_j^{n+1})^*$, if $(j = 1, \cdots, n)$ and $(1, \cdots, 1)^*$ if $j = n + 1$, is nonsingular.

Proof

Assume to the contrary that the above matrix is singular and take $u \neq 0$ in its null space. Then

$$
\begin{pmatrix} A_1^1 & \cdots & A_n^1 & 1 \\ \vdots & & \vdots & \vdots \\ A_1^{n+1} & \cdots & A_n^{n+1} & 1 \end{pmatrix} \begin{pmatrix} u_1 \\ \vdots \\ u_{n+1} \end{pmatrix} = \begin{pmatrix} 0 \\ \vdots \\ 0 \end{pmatrix}
$$

Thus,

$$
\sum_{j=1}^{n} u_j A_j^i = -u_{n+1}, \qquad (1 \leq i \leq n + 1).
$$

If $u_{n+1} = 0$, the n-dimensionality of S is violated. If $u_{n+1} \neq 0$, either $\pm u$ satisfies the system of inequalities $[A^i, x] < 0, (1 \leq i \leq n + 1)$. By Section A-1, this implies that $0 \notin H(S)$.   Q.E.D.

Exchange Theorem

Assume that the set $\{A^i : 1 \leq i \leq n + 2\}$ satisfies the Haar condition and that $0 \in H\{A^i : 1 \leq i \leq n + 1\}$. There exists a unique index j in $\{1, \cdots, n + 1\}$ such that $0 \in H\{A^i : 1 \leq i \leq n + 2, i \neq j\}$.

Proof

Since $0 \in H\{A^1, \cdots, A^{n+1}\}$, there exists numbers $u_1, \cdots, u_{n+1}$ such that

$$
\sum_{i=1}^{n+1} u_i A^i = 0, \qquad \sum_{i=1}^{n+1} u_i = 1, \qquad \text{and } u_i \geq 0, \quad (1 \leq i \leq n + 1).
$$

By the above lemma, these numbers are unique; by the Haar condition, all are positive. Again by the above lemma, there exists a unique set of $n + 1$ numbers, $v_1, \cdots, v_{n+1}$ such that

$$
A^{n+2} = \sum_{i=1}^{n+1} v_i A^i \qquad \text{with } \sum_{i=1}^{n+1} v_i = 1.
$$

For any $j \leq n + 1$ we have

$$0 = A^{n+2} - \sum_{i=1}^{n+1} v_i A^i = A^{n+2} - v_j A^j - \sum_{i \neq j} v_i A^i$$

$$= A^{n+2} + v_j \sum_{i \neq j} \frac{u_i}{u_j} A^i - \sum_{i \neq j} v_i A^i$$

$$= A^{n+2} + \sum_{i \neq j} \left(\frac{v_j u_i}{u_j} - v_i \right) A^i.$$

If $v_j u_i / v_j \geq v_i$ for all i, we can represent the 0-vector as a convex combination of points out of $\{A^i : 1 \leq i \leq n + 2, i \neq j\} = \mathscr{A}$. Choose j so that $v_j / u_j \geq v_i / u_i$ for all $(i = 1, 2, \cdots, n + 1)$. Indeed, for this value of j, $v_j / u_j > v_i / u_i$ for all $i \neq j$ because, otherwise, the Haar condition would be violated. Finally, we divide both sides of the equation

$$0 = A^{n+2} + \sum_{\substack{i=1 \\ i \neq j}}^{n+1} \left(\frac{v_j u_i}{u_j} - v_i \right) A^i$$

by an appropriate positive number, and we obtain the desired representation of 0 as a convex combination out of $\mathscr{A}$.   Q.E.D.

Let $\{1, \cdots, k\} = I$ and $\{k + 1, \cdots, m\} = J$. Let $A^i, b^i \, i \in I \cup J$ be points of E_n and E_1, respectively. Set

$$R^i(x) = [A^i, x] - b^i \qquad \text{for } i \in I \cup J.$$

We consider next the characterization of points that minimize $F(x) = \max\{R^i(x) : i \in I\}$ on the set $K = \{x \in E_n : R^i(x) \leq 0 \text{ for all } i \in J\}$, if such exist.

Theorem

Assume K is nonempty. F achieves a minimum at z on K only if $0 \in H(S_1 \cup S_2)$, where

$$S_1 = \{A^i : i \in I \quad \text{and} \quad R^i(z) = F(z)\}$$

and

$$S_2 = \{A^i : i \in J \quad \text{and} \quad R^i(z) = 0\}.$$

Conversely, the above condition is sufficient, provided $0 \notin H\{S_2\}$.

Proof

Assume that z is a solution. Let $\bar{I} \subset I$ and $\bar{J} \subset J$ denote the indices for which $R^i(z) = F(z)$, and $R^i(z) = 0$, respectively. Observe that $\bar{J}$ may be empty, but $\tilde{I}$ is not. If $0 \notin H(S)$, then—by (b) of the exercise of Section B-1, page 70—since the convex hull of a finite point set is closed, the system

$$\begin{aligned}[A^i, x] &< 0 \qquad i \in \tilde{I} \\ [A^i, x] &\le 0 \qquad i \in \tilde{J}\end{aligned}\Bigg\}$$

is consistent, with solution, say, at x. For all positive λ, the vector $z + \lambda x$ has the property that $[A^i, z + \lambda x] - b^i < F(z)$ for $i \in \tilde{I}$, and $[A^i, z - \lambda x] - b^i \le 0$ for $i \in \tilde{J}$. Moreover, if $i \in I \sim \tilde{I}$, $R^i(z) < F(z)$, and if $i \in J \sim \tilde{J}$, $R^i(z) < 0$. It follows, therefore, that these inequalities remain valid at $z + \lambda x$ for small positive λ because of the continuity of the functions R^i, $(1 \le i \le m)$. Therefore, we see that z is not a minimizer of F on K, contrary to the hypothesis.

For the sufficiency, assume z to be given such that $0 \in H(S_1 \cup S_2)$. Assume that z is not extremal. Then, for some $y \in K$, $R^i(y) < F(z)$ for $i \in \tilde{I}$ and $R^i(y) \le 0$ for $i \in \tilde{J}$. Thus, $[A^i, y - z] < 0$ for $i \in \tilde{I}$ and $[A^i, y - z] \le 0$ for $i \in \tilde{J}$. Since $0 \notin H(S_2)$, we have by (a) of the exercise on p. 70 that

$$0 \notin H(S_1 \cup S_2).$$

Corollary (characterization theorem)

Assume that K is nonempty and that $\{A^i : i \in I \cup J\}$ satisfies the Haar condition. A point z in K is a unique minimizer of F on K iff there exists a subset $\tilde{I} \subset I$ and $\tilde{J} \subset J$ such that $\tilde{I} \cup \tilde{J}$ contains exactly $n + 1$ points and $0 \in H\{A^i : i \in \tilde{I} \cup \tilde{J}\}$.

Proof

The proof of necessity is partially completed in the preceding theorem. Because of the Haar condition and because $0 \in H(S_1 \cup S_2)$, $S_1 \cup S_2$ must contain at least $n + 1$ points. By Caratheodori's lemma, a subset S' of $S_1 \cup S_2$ can be chosen with precisely $n + 1$ points such that $0 \in H(S')$. If $S' \cap S_1$ is empty, we can rectify by the Exchange theorem. Conversely, since $\tilde{I}$ cannot be empty, it follows that $\tilde{J}$ contains at most n points. Therefore, because of the Haar condition, it follows that $0 \notin H(A^i : i \in \tilde{J})$. Thus, the sufficiency follows as in the theorem above.

We now show that if z is an extremal, then z is unique. Consider the system

$$\begin{aligned}(a) &\qquad\qquad [A^i, x] < 0 \qquad i \in \tilde{I} \\ (b) &\qquad\qquad [A^i, x] \le 0 \qquad i \in \tilde{J}\end{aligned}\Bigg\}$$

Take any $x \neq 0$ satisfying (b). Such x exist because $0 \notin H\{A^i: i \in \bar{J}\}$. We now prove that for some $i \in \bar{I}$, $[A^i, x] > 0$.

To see this, observe that $[A^i, x] \not\equiv 0$, $i \in I \cup J$. Therefore, since

$$\Sigma \lambda_i [A^i, x] = 0 \quad \text{and} \quad \lambda_i > 0 \quad \text{for all} \quad i \in \bar{I} \cup \bar{J},$$

it follows that $[A^i, x]$ must assume positive values on $\bar{I}$ and negative values on $\bar{J}$. Take $y \in K$ distinct from z. Then $[A^i, y] - b^i \leq 0$ $i \in \bar{J}$. Hence, $[A^i, y - z] \leq 0$ for $i \in \bar{J}$ and $[A^i, y - z] > 0$ for some $i \in \bar{I}$. This shows that $[A^i, y] - b^i > [A^i, z] - b^i = F(z)$.   Q.E.D.

EXERCISES

1. Let $[A^i, z] \leq b^i$ $(i = 1, \cdots, m)$. Assume that

$$T = \{A^i: 1 \leq i \leq k < m\}$$

satisfies the Haar condition and $0 \in H(T)$. Then $k > n$. Assume that $[A^i, z] = b^i$, $(i = 1, \cdots, k)$. Prove that the set $K = \{x: [A^i, z] \leq b^i, 1 \leq i \leq m\}$ contains only the point z.

2. Let $S = \{A^i: i = 1 \leq i \leq n + 2\}$ be a subset of E_n satisfying $0 \in H(S)$ and the Haar condition. There are just two distinct subsets of S, say S_1 and S_2, which contain $n + 1$ points and are such that $0 \in H(S_i)$, $i = 1, 2$.

The characterization theorem, p. 73, provides us with a test to apply to the subsets with $n + 1$ points of $\{A^i: i \in I \cup J\} = S$. One such subset, say $\bar{I} \cup \bar{J}$, satisfies these tests. The extremal x can be found by solving the linear system of $n + 1$ equations,

$$\left. \begin{array}{ll} [A^i, x] - F(x) = b^i & i \in \bar{I} \\ [A^i, x] - b^i = 0 & i \in \bar{J} \end{array} \right\},$$

for the $n + 1$ unknowns $(x_1, x_2, \cdots, x_n, F(x))$. (See the Exercise 1 below.)

The labor of testing all the subsets of S containing $n + 1$ points, however, is enormous for large systems. The algorithm we now give finds $\bar{I} \cup \bar{J}$ in a systematic way.

EXERCISES

1. Show that if $0 \in H\{A^i: i \in \bar{I} \cup \bar{J}\}$, then the above system of linear equations has a unique solution.

2. Let $S = \{A^i: 1 \leq i \leq n + 1\}$. Assume that S is n-dimensional. Suppose that

$$0 \in \sum_{i=1}^{n+1} \lambda_i A^i \quad \text{with} \quad \lambda_i > 0 \quad \text{and} \quad \Sigma \lambda_i = 1.$$

Show that 0 belongs to the interior of $H(S)$ and that S satisfies Haar conditions. *Hint*: Set up the system of linear equations for $0 \in H(S)$ and use Cramer's rule and the lemma at the beginning of this section.

Remark

The above problem of minimizing F on K contains the linear programming problem of minimizing a linear function on a polytope. This is the case when I is the singleton $\{1\}$.

Algorithm

Set $F(x) = \max\{R^i(x): i \in I = \{1, 2, \cdots, k\}\}$ and

$$G(x) = \max\{R^i(x): i \in J = \{k + 1, \cdots, m\}\}.$$

Let $K = \{x: G(x) \le 0\}$. Assume that $\{A^1, \cdots, A^m\}$ satisfies the Haar condition. Assume that there exists a subset $\bar{I} \subset I$ and $\bar{J} \subset J$ such that

(i) $\bar{I} \cup \bar{J}$ has exactly $n + 1$ points,
(ii) $0 \in H\{A^i: i \in \bar{I} \cup \bar{J}\}$,
(iii) $\bar{I}$ is nonempty.

Then there exists a number μ and a vector x such that

(iv) $R^i(x) = \mu$ for $i \in \bar{I}$,
(v) $R^i(x) = 0$ for $i \in \bar{J}$.

If $x \in K$ and $F(x) = \mu$, then $F(x) < F(y)$ for all $y \neq x$ in K. If $G(x) > 0$, select $p \in J$ so that $R^p(x) > 0$. If $G(x) \le 0$ and $F(x) > \mu$, select $p \in I$ so that $R^p(x) > \mu$. Choose $q \in \bar{I} \cup \bar{J}$ so that $0 \in H\{A^i: i \in \bar{I} \cup \bar{J} \cup \{p\} \sim \{q\} = \bar{I}' \cup \bar{J}'\}$. Then the set $\bar{I}' \cup \bar{J}'$ has the properties (i) to (iii) above, and points x', μ', exist, which satisfy (iv) and (v). By a finite repetition of the above process, we obtain an extremal satisfying (iv) and (v). The sequence $\{\mu, \mu', \cdots\}$ generated by this process forms a monotone-increasing sequence.

Proof

To obtain the point x and μ, we have, assuming conditions (i), (ii), and (iii) and the result of exercise 1, above, that the system of linear equations

$$\left. \begin{array}{ll} [A^i, x] - \mu = b^i & i \in \bar{I} \\ [A^i, x] = b^i & i \in \bar{J} \end{array} \right\}$$

has a unique solution for the pair (x, μ). By the characterization theorem, if $x \in K$ and $F(x) = \mu$, then x is an extremal.

For the remaining cases it is clear by the exchange lemma that $\bar{I}' \cup \bar{J}'$

satisfies (i) and (ii). We now show that (iii) is satisfied as well. If this is not the case, there is but one point in the set $\bar{I}$ and $p \in J$. Select $y \neq x$ in K. Then $[A^i, y] \leq b^i$ for $i \in J$. For $i \in \bar{J}$, $[A^i, x] = b^i$ and $[A^i, y - x] \leq 0$. Since

$$[A^p, x] - b^p > 0, \quad [A^p, y - x] < 0.$$

Hence, $0 \notin H\{A^i : i \in \bar{I}' \cup \bar{J}'\}$, contradicting property (ii). Therefore, it follows that there is no difficulty in obtaining an x' and μ' satisfying

$$\left.\begin{array}{ll} [A^i, x'] - b^i = \mu' & i \in \bar{I}' \\ [A^i, x'] - b^i = 0 & i \in \bar{J}' \end{array}\right\}.$$

Since

$$\left.\begin{array}{ll} [A^i, x] - b^i = \mu & i \in \bar{I} \\ [A^i, x] - b^i = 0 & i \in \bar{J} \end{array}\right\},$$

we have that

$$\left.\begin{array}{ll} [A^i, x' - x] = \mu' - \mu & i \in \bar{I}' \cap \bar{I} \\ [A^i, x' - x] = 0 & i \in \bar{J}' \cap \bar{J} \end{array}\right\}.$$

If $p \in I$, we have $[A^p, x] - b^p > \mu$, while $[A^p, x'] - b^p = \mu'$. If $p \in J$, we have $[A^p, x] - b^p > 0$, while $[A^p, x'] - b^p = 0$, whence

$$\left.\begin{array}{ll} [A^p, x' - x] < \mu' - \mu & \text{if } p \in I \\ [A^p, x' - x] < 0 & \text{if } p \in J \end{array}\right\}.$$

If $\mu' - \mu \leq 0$, we verify that $0 \notin H\{A^i : i \in \bar{I}' \cup \bar{J}'\}$. Hence, $\mu' - \mu > 0$. Thus, the sequence $\{\mu, \mu', \cdots\}$ that is generated by the algorithm is strictly monotone. To each set $\bar{I} \cup \bar{J}$ there corresponds a unique pair (x, μ) which satisfies the system given by (iv) and (v). The number of distinct sets of the form $\bar{I} \cup \bar{J}$ being finite, the set of distinct pairs (x, μ) are finite. Thus, the sequence $\{(x, \mu), (x', \mu') \cdots\}$ must terminate, since no pair can repeat. Clearly, the termination must occur with a pair, say, $(\bar{x}, \bar{\mu})$ such that $\bar{x} \in K$ and $F(\bar{x}) = \bar{\mu}$.

EXERCISE

Is $\bar{I}' \cap \bar{I}$ always nonempty?

Remark

If K is empty, the algorithm will indicate this when (iii) does not hold at some juncture of the algorithm.

Proof

If on the contrary, (iii) always obtains, then by the above proof, the algorithm will generate a finite sequence $\{\mu, \mu', \cdots\}$. On the other hand, because K is empty, the point p will always belong to J and the algorithm will never terminate. Thus, the sequence $\{\mu, \mu', \cdots\}$ does not terminate—a contradiction.

Remark

Realistic bounds for the total number of cycles for the preceding algorithm are not known, except in special cases.

EXERCISES

1. Give an example in E_2 where F does not achieve a minimum on K.

2. Assume that K is not empty and that $\{A^i : i \in I \cup J\}$ satisfies the Haar conditions. Prove that the following statements are equivalent:
   (a) There exists a subset $\bar{I}$ of I and $\bar{J}$ of J such that $\bar{I} \cup \bar{J}$ satisfies conditions (i), (ii), and (iii) of the algorithm.
   (b) F achieves a minimum on K.

B-3. Implementation of the Algorithm

We are interested in employing the algorithm in Section B-2 to situations where the Haar condition is not satisfied. The term *degenerate* is often used to characterize points set in E_n that do not possess the Haar condition. This is typical of terms used for situations that are troublesome.

Let us assume that the set $\{A^1, \cdots, A^m\}$ is n-dimensional and does not satisfy the Haar condition. Assume further that the set $\{A^i : 1 \leq i \leq n + 1\}$ does satisfy the Haar condition and moreover contains 0 in its convex hull. Such a set can be obtained by the starting process below. Suppose now that we apply the algorithm of Section B-2, and that at the first cycle, the point A^{n+1} is to be exchanged with one of the elements in $\{A^i : 1 \leq i \leq n + 1\}$. Let

$$A^{n+2} = \sum_{i=1}^{n+1} v_i A^i \quad \text{and} \quad 0 = \sum_{i=1}^{n+1} u_i A^i, \ \sum_{i=1}^{n+1} u_i = 1, \qquad u_i > 0.$$

We have, as in Section B-2, that

$$0 = A^{n+1} + \sum_{\substack{i=1 \\ i \neq j}}^{n+1} \left(\frac{v_j u_i}{u_j} - v_i \right) A^i$$

for any j, $1 \leq j \leq n + 1$. Choose j as before so that $v_j/u_j \geq v_i/u_i$ for all $i \neq j$. If equality holds for some $i = i'$, then a unique exchange cannot be made; moreover, the new index set will contain fewer than $n + 1$ points because $(v_j u'_{i}/u_j) - v_{i'} = 0$. We shall rectify this situation by "perturbing" A^{n+1}. Choose a positive number ε. Set

$$\bar{A}^{n+2} = A^{n+2} + \sum_{i=1}^{n+1} i\varepsilon u_i \, \text{sgn}(j - i)A^i,$$

where $\text{sgn}(x) = 1$ if $x \geq 0$ and $\text{sgn}(x) = -1$ otherwise. Thus,

$$\bar{A}^{n+2} = \sum_{i=1}^{n+1} \bar{v}_i A^i,$$

where

$$\bar{v}_i = v_i + i\varepsilon u_i \, \text{sgn}(j - i), \qquad 1 \leq i \leq n + 1,$$

and as above,

$$0 = \bar{A}^{n+2} + \sum_{\substack{i=1 \\ i \neq j}}^{n+1} \left(\frac{\bar{v}_j u_i}{u_j} - \bar{v}_i\right)A^i.$$

We now examine the coefficients

$$\left(\frac{\bar{v}_j u_i}{u_j} - \bar{v}_i\right) = \alpha_i.$$

If $i = j$, clearly $\alpha_j = 0$; otherwise:

$$\alpha_i = \frac{v_j u_i}{u_j} - v_i + \varepsilon u_i(j - i \, \text{sgn}(j - i)).$$

Thus,

$$\left.
\begin{aligned}
\alpha_i &= \frac{v_j u_i}{u_j} - v_i + \varepsilon u_i(j - i) && \text{if } \ j > i \\[2mm]
&= \frac{v_j u_i}{u_j} - v_i + \varepsilon u_i(j + i) && \text{if } \ j < i
\end{aligned}
\right\}.$$

It follows that $\alpha_i > 0$ for all $i \neq j$. Thus, a unique exchange can be made, and a new set $\{A^1, \cdots A^{j-1}, A^{j+1}, \cdots \bar{A}^{n+2}\} = S$ can be found with the property that S satisfies the Haar condition (see Exercise 2, Section B-2, p. 74), and $0 \in H(S)$.

Our object now is to obtain a new problem, *the perturbed problem*, by small perturbations of some of the points of $\{A^1, \cdots, A^m\}$. The perturbations will be made in such a way that our algorithm can be used to solve the perturbed problem.

In what follows, $\hat{A}^i$ will indicate elements that are either A^i or perturbations $\bar{A}^i$ of A^i. Subsets I, of $n + 1$ indices with the property that the set $\{A^i : i \in I\}$ contains 0 in its convex hull and satisfies the Haar condition, will be said to satisfy property P. We have seen that given a subset I_0 with property P and a point $\hat{A}^k$, $k \notin I_0$, a new subset I_1 can be obtained which contains the index k and other n-indices from I_0 such that the set $\{A^i : i \in I_1 \cap I_0\} \cup \{\bar{\hat{A}}^k\}$ has property P and $\bar{\hat{A}}^k$ is arbitrarily close to $\hat{A}^k$.

Suppose I is a subset satisfying property P. Then any subset of elements of $\{\hat{A}^i : i \in I\}$ can be slightly "perturbed," and provided these perturbations are not too large, the resulting set still has property P. (Why?)

Now assume that subsets $I_0, I_1, \cdots, I.$ have been constructed, all possessing property P. The set I_{s+1} is constructed as follows: The point $\hat{A}^k$ is furnished by the algorithm. If no perturbation of $\hat{A}^k$ is required, I_{s+1} is obtained simply by the use of the exchange theorem. If a perturbation of $\hat{A}^k$ is required, an exchange will be made by the method above; however, the ε used must be sufficiently small so that not only does I_{s+1} have property P, but also so that, if $\hat{A}^k$ belongs to any of the sets $I_0, \cdots, I_s$, property P will not be sacrificed when $\hat{A}^k$ is perturbed in these sets.

Consider the sequence $\{I_0, I_1, \cdots\}$. Either this sequence terminates or, if not, there is "*cycling*"; that is, a finite subsequence, say $\{I_{i_0}, I_{i_1}, \cdots I_{i_r}\}$ is repeated infinitely often. Let $\{I_{i_0}^k, I_{i_1}^k, \cdots I_{i_r}^k\}$, $(k = 0, 1, 2, \cdots)$, denote the kth repetition of this sequence. If $k \geq 1$, no perturbations need be made in order to construct the sequence $\{I_{i_0}^k, \cdots I_{i_r}^k\}$.

Because the number μ generated by the algorithm strictly increases as we proceed from I_j^k to I_{j+1}^k, we arrive at a contradiction. Therefore, the sequence $\{I_0, I_1, \cdots\}$ terminates with the solution of a perturbed system; namely, we have obtained a point, say z^*, which minimizes

$$F^*(x) = \max_{i \in I}\{[\hat{A}^i, x] - b^i\}$$

on the set $K^* = \{x : [\hat{A}^i, x] - b^i \leq 0 \text{ for all } i \in J\}$, where the $\hat{A}^i$ are arbitrarily close to A^i, $(1 \leq i \leq m)$.

EXERCISES

1. The problem of polyhedral convex programming may be stated as:

Minimize $F(x)$ subject to $G(x) \leq 0$ (See the algorithm above of Section B-2.) Suppose z^* minimizes F^* on K^* and z minimizes F on K. Show that if $\delta = \max\{\|\hat{A}^i - A^i\| \|z^*\| : i \in I \cup J\}$, then

$$|F^*(z^*) - F(z)| < \delta \quad \text{and} \quad G(z^*) \leq \delta.$$

2. Assume 0 belongs to the interior of $H\{A^i : i \in I \cup J\}$. Show that $\{x \in K : F(x) \leq M\}$ is bounded for all M.

Hints:

$$\left.\begin{array}{ll} \left[A^i, \dfrac{x}{\|x\|}\right]\|x\| \leq \max_{i \in I} b^i + M & i \in I \\[2em] \left[A^i, \dfrac{x}{\|x\|}\right]\|x\| \leq \max_{i \in J} b^i & i \in J \end{array}\right\}$$

and

$$\min_{\|x\| = 1} \max_i [A^i, x] \geq q > 0.$$

To prove this latter statement, observe that since $0 \in \text{int } H\{A^i : i \in I \cup J\}$, every representation of 0 as $0 = \Sigma \lambda_i A^i$ has at least $n + 1$ nonzero λ_i. Hence, for every $x \neq 0$ and some i, $[A^i, x] > 0$.

3. Assume the hypothesis of Exercise 2. Show that if the elements $\hat{A}^i$ tend to A^i, the solution z^* tends to z.

The Starting Process

In general, a subset $\bar{I} \cup \bar{J}$ satisfying the conditions (i), (ii), and (iii) required by the algorithm in Section B-2 will not be known. When a subset $\bar{I} \cup \bar{J}$ satisfying (i), (ii), and (iii) exists, the algorithm shows that there exists a point z in K minimizing F. A priori, however, we have no way of checking whether a set with the desired properties actually exists, or equivalently, whether F actually achieves a minimum. Nevertheless, the algorithm to be described will construct a subset with the desired property if such exists, or will otherwise indicate that F does not achieve a minimum.

It is easy to see that in order to obtain our starting set, we may confine our attention to a simpler problem; namely, minimizing the function

$$G(x) = \max\{[A^i, x] - b^i : i \in I \cup J\}.$$

For if $0 \notin H\{A^i : i \in I \cup J\}$ and x satisfies $[A^i, x] < 0$, $i \in I \cup J$, then $G(\lambda x) \to -\infty$ as $\lambda \to \infty$. Conversely, if $0 \in H\{A^i : i \in I \cup J\}$, G achieves a minimum, and the minimum will be achieved on some index set $\bar{I} \cup \bar{J}$ satisfying conditions (i), (ii) of the algorithm. In the event $(\bar{I} \cup \bar{J}) \cap I$ is empty, the situation can be remedied by the exchange theorem. Let $\hat{I} \cup \hat{J}$ be a subset of n points of $I \cup J$ such that $\{A^i : i \in \hat{I} \cup \hat{J}\}$ spans E_n.

Set $A^0 = -\Sigma A^i$, where i ranges over $\hat{I} \cup \hat{J}$ and set $S = \{A^i : i \in \hat{I} \cup \hat{J} \cup \{0\}\}$. Then S satisfies the Haar condition and conditions (i), (ii), and (iii) of the algorithm in Section B-2.

Clearly, conditions (i), (ii), and (iii) are satisfied; moreover, it is almost immediate that the Haar condition is also satisfied. For if on the contrary some proper subset S' of S were linearly dependent, then

$$0 = \sum_{i \in S'} u_i A^i, \qquad \text{with all } u_i \text{ nonzero.}$$

Clearly, $A^0 \in S'$, and replacing A^0 by $-\Sigma A^i$ leads to a contradiction. We now use the algorithm to construct the required subset of $I \cup J$ satisfying conditions (i), (ii), and (iii).

Starting Algorithm

Set $b^0 = 1$ and $G^*(x) = \max\{[A^i, x] - b^i : i \in I \cup J \cup \{0\}\}$ with some $b^i \neq 0$ for $i \neq 0$. Apply the algorithm of the above, Section B-2, taking $\bar{I} \cup \hat{J} \cup \{0\} = S$ to start and using the perturbation procedure if necessary. Then the following possibilities arise:

(i)   At some juncture of the algorithm, an index set $\bar{I} \cup \bar{J}$ is found not containing 0. This is the desired starting set.
(ii) At the solution z, $R^0(z) = G^*(z)$.

If case (ii) occurs, replace b^0 by $10b^0$ and repeat. There are then two possibilities. Within a finite number of repetitions, either case (i) occurs or eventually $G^*(z) < \max_i -b^i$. In the latter event, the function F does not achieve a minimum on K.

Proof

Clearly, G^* achieves a minimum for each b^0. Let $z = z(b_0)$ and assume $G(z) \leq G^*(x)$ for all $x \in E_n$. If G achieves a minimum at, say, z', then $G^*(z') \geq G(z')$. If $R^0(z') < G(z')$, then $G^*(z') = G(z') = G(z)$. Clearly, for b^0 sufficiently large, $R^0(z') < G(z')$. This shows that case (i) will eventually occur if

$$0 \in H\{A^i : I \cup J\}.$$

Moreover, if $0 \in H\{A^i : i \in I \cup J\}$,

$$G(x) = \max\{[A^i, x] - b^i\} \geq \max[A^i, x] - \max b^i \geq -\max b^i.$$

Thus, if $G(x) < -\max b^i$, $0 \notin H\{A^i : I \cup J\}$ and F does not achieve a minimum on K.

Now assume that $0 \notin H\{A^i : i \in I \cup J\}$. We show that for b_0 sufficiently large, $G^*(z) < -\max b^i$. Observe that $G^*(z)$ is the smallest number for which the system

$$
\left.
\begin{aligned}
[A^i, x] - b^i \leq G^*(z) \qquad i \in I \cup J \\
[A^0, x] - b^0 \leq G^*(z)
\end{aligned}
\right\}
$$

is consistent. Take a number $M < -\max b^i$. We show that $M > G^*(z)$. Since $0 \notin H\{A^i: i \in I \cup J\}$, there exists x such that $[A^i, x] \leq M + b^i$ for all $i \in I \cup J$. Moreover, for some b^0 sufficiently large, $[A^0, x] \leq M + b^0$. It follows, therefore, that for b^0 sufficiently large, $G(z) \leq G^*(z) \leq M < -\max b^i$.

Q.E.D.

SECTION C. INFINITE CONVEX PROGRAMMING

C-1.   Nonpolyhedral Convex Programming I

We now consider a possible generalization of the algorithm of Section B-2 by replacing finite sets by compact sets and retaining Haar conditions.

Specifically, let Θ denote a compact metric space, A and b continuous maps from Θ to E_n and E_1, respectively. Assume that $A(\Theta)$ satisfies Haar conditions, and that $\Theta = \Theta' \cup \Theta''$, with Θ' and Θ'' closed and disjoint. Let $R(A(\theta), x) = [A(\theta), x] - b(\theta)$, $F(x) = \max\{R((A(\theta), x): \theta \in \Theta'\}$, $G(x) = \max\{R(A(\theta), x): \theta \in \Theta''\}$ and $K = \{x \in E_n: G(x) \leq 0\}$. We shall assume that each x and a given $\varepsilon > 0$, there can be found $\theta' \in \Theta'$ and $\theta'' \in \Theta''$ such that $F(x) - R(A(\theta'), x) < \varepsilon$ and $G(x) - R(A(\theta''), x) < \varepsilon$.

Algorithm

Assume the above hypotheses. Assume that there exists a set $\Theta_1 \subset \Theta$ with the properties:

(i) Θ_1 contains $n + 1$ points.
(ii) $0 \in H\{A(\theta): \theta \in \Theta_1\}$.
(iii) $\Theta_1 \cap \Theta'$ nonempty.

For every $k = 1, 2, 3$ there exists Θ_k satisfying (i), (ii), and (iii). Choose x^k and μ_k to satisfy:

(iv) $R(A(\theta), x^k) = \mu_k$, $\theta \in \Theta_k \cap \Theta'$.
(v) $R(A(\theta), x^k) = 0$, $\theta \in \Theta_k \cap \Theta''$.

If $G(x^k) \leq 0$ and $F(x^k) = \mu_k$, x^k minimizes F on K. If this is not the case, and $F(x^k) - \mu_k \geq G(x^k)$, set $\gamma_k = F(x^k) - \mu_k$ and choose $\theta' \in \Theta'$ such that

$$R(A(\theta'), x^k) - \mu_k \geq \gamma_k/2;$$

otherwise, set $\gamma_k = G(x^k)$ and choose $\theta' \in \Theta''$ such that $R(A(\theta'), x^k) \geq \gamma_k/2$. In either case choose $\theta'' \in \Theta_k$ so that $0 \in H\{A(\Theta_{k+1})\}$, where

$$\Theta_{k+1} = (\Theta_k \cup \{\theta'\} \sim \{\Theta''\}).$$

Replace Θ_k by Θ_{k+1} and calculate x^{k+1}, μ_{k+1} from (iv), (v). Then:

(a) $\{\mu_k\}$ converges upward to $\inf\{F(x): x \in K\} = \mu$;
(b) $\{x^k\}$ converges to z in K, and $F(z) = \mu$;
(c) z is unique.

Proof

The existence of Θ_k satisfying (i), (ii), and (iii) for every $k = 2, 3, 4, \cdots$ follows from Section B-2. We also recall from Section B-2 that x^k and μ_k are uniquely determined.

We now show that if $G(x^k) \leq 0$ and $F(x^k) = \mu_k$, then x^k minimizes F on K. If not, then for some $z \in K$, $F(z) < \mu_k$. Then, since, for example,

$$\max\{[A(\theta), z] - b(\theta): \theta \in \Theta_k \cap \Theta'\} \leq F(z),$$

we have

$$[A(\theta), z - x^k] < 0 \qquad \text{for} \quad \theta \in \Theta_k \cap \Theta',$$

$$[A(\theta), z - x^k] \leq 0 \qquad \text{for} \quad \theta \in \Theta_k \cap \Theta'',$$

where $\Theta_k \cap \Theta''$ is possibly empty. By the familiar argument of Section B-2, $0 \notin H\{A(\theta): \theta \in \Theta_k)$, a contradiction.

We now prove (a). Let $F_k(x) = \max\{R(A(\theta), x): \theta \in \Theta_k \cap \Theta')$. Clearly, $F_k(x) \leq F(x)$ for all x. Let $K_k = \{x: R(A(\theta), x) \leq 0: \theta \in \Theta_k \cap \Theta''\}$. Observe that μ_k is the minimum of F_k on the set K_k and that $K \subset K_k$. Thus,

$$\inf\{F_k(x) : x \in K_k\} = \mu_k \leq \inf\{F(x): x \in K_k\} \leq \mu.$$

Let $\hat{\Theta}$ be any subset of Θ such that $0 \in H\{A(\theta): \theta \in \hat{\Theta}\}$. Let α denote the infimum of the smallest coefficient, say $\bar{\lambda}$, of the sum $0 = \{\sum \lambda(\theta)A(\theta): \theta \in \Theta\}$ over all possible subsets Θ. We show that $\alpha > 0$. Let $\{\hat{\Theta}_i\}$ be a sequence of subsets of Θ, satisfying (i) and (ii). Thus, we may consider $\hat{\Theta}_i$ to be a point in the $n + 1$-fold Cartesian product of Θ, which is compact. Let the sequence $\hat{\Theta}_i$ have the property that $\{\bar{\lambda}_i\}$ converges downward to α, and extract a subsequence converging to a cluster point $\overline{\Theta}$. Since properties (i) and (ii) hold for each member of the sequence, and because λ is a continuous function of $\hat{\Theta}$, the properties (i) and (ii) hold in the limit. But, because of property (ii) and the Haar conditions, $\alpha > 0$. At this juncture we shall obtain a lower bound for the increase in μ_k as we proceed from Θ_k to Θ_{k+1}. We calculate that

$$[A(\theta), x^k - x^{k+1}] = \mu_k - \mu_{k+1} \qquad \theta \in \Theta_k \cap \Theta_{k+1} \cap \Theta',$$

$$[A(\theta), x^k - x^{k+1}] = 0 \qquad \theta \in \Theta_k \cap \Theta_{k+1} \cap \Theta'',$$

and

$$[A(\theta'), x^k - x^{k+1}] \geq \mu_k - \mu_{k+1} + \frac{\gamma_k}{2} \qquad \text{if} \quad \theta' \in \Theta' \cap \Theta_{k+1}$$

or

$$[A(\theta'), x^k - x^{k+1}] \geq \frac{\gamma_k}{2} \qquad \text{if} \quad \theta' \in \Theta'' \cap \Theta_{k+1}.$$

Since $\Theta_{k+1} = \Theta_k \cap \Theta_{k+1} \cup \{\theta'\}$, and $0 \in H\{A(\theta): \theta \in \Theta_{k+1}\}$, we have, after multiplying by $\lambda(\theta)$ and summing,

$$0 \geq (\mu_k - \mu_{k+1}) \sum \lambda(\theta) + \lambda(\theta')\left(\mu_k - \mu_{k+1} + \frac{\gamma_k}{2}\right) \qquad \text{if} \quad \theta' \in \Theta'$$

and

$$(\mu_k - \mu_{k+1}) \sum \lambda(\theta) + \lambda(\theta')\frac{\gamma_k}{2} \leq 0 \qquad \text{if} \quad \theta' \in \Theta'',$$

the summation in these equations being over $\Theta_k \cap \Theta_{k+1} \cap \Theta'$. Thus, in both cases,

$$\mu_{k+1} - \mu_k > \lambda(\theta')\frac{\gamma_k}{2} \geq \frac{\alpha\gamma_k}{2}.$$

If $\gamma_k \nrightarrow 0$, then there is a subsequence γ_k and an $\varepsilon > 0$ such that $\gamma_k > \varepsilon$. Hence, $\mu_k \to \infty$, contradicting that $\mu_k \leq \mu$. Since the numbers μ_k are bounded above by μ, they converge, say, to μ_0. If $\mu_0 \neq \mu$, we contradict that $\gamma_k \to 0$.

Next we prove that the sequence $\{x_k\}$ is bounded. For each k,

$$[A(\theta), x^k] - b(\theta) = \mu_k \qquad \text{for} \quad \theta \in \Theta_k \cap \Theta'$$

and

$$[A(\theta), x^k] - b(\theta) = 0 \qquad \text{for} \quad \theta \in \Theta_k \cap \Theta''.$$

We have that

$$\max\{[A(\theta), x_k]: \theta \in \Theta_k\} \geq \inf_{x \neq 0} \max\left\{\frac{[A(\theta), x]}{\|x\|}: \theta \in \Theta_k\right\}.$$

Since $0 \in H\{A(\theta): \theta \in \Theta_k\}$, $\max[A(\theta), (x/\|x\|)] \geq 0$. Indeed, strict inequality obtains because of the Haar conditions. Thus,

$$\max\left[A(\theta), \frac{x}{\|x\|}\right] \geq \delta > 0 \qquad \text{for all } x \neq 0.$$

Also,

$$\max\{[A(\theta), x^k]: \theta \in \Theta_k] \geq \delta\|x^k\|,$$

and

$$\max \frac{\{b(\theta) + \mu: \theta \in \Theta'\}}{\delta} \geq \|x_k\|,$$

showing that $\{x^k\}$ is bounded. If z is a cluster point of $\{x^k\}$, clearly $G(z) \leq 0$ and $F(z) = \mu$. A further subsequence of $\{x^k\}$ may be extracted such that the $(n + 1)$-tuples Θ_k, $k = 1, 2, 3, \cdots$, have a cluster point $\overline{\Theta}$ in the $(n + 1)$-fold Cartesian product of Θ. Since $0 \in H\{A(\Theta_k)\}$ for each k, we argue as before to show that $0 \in H\{A(\Theta)\}$, and thus $\overline{\Theta}$ contains $n + 1$ distinct elements. By Section B-2, the solution of the system below for the pair (x, λ) exists and is unique.

$$\left.\begin{aligned} R(A(\theta), x) = \lambda &\qquad \theta \in \overline{\Theta} \cap \Theta' \\ R(A(\theta), x) = 0 &\qquad \theta \in \overline{\Theta} \cap \Theta'' \end{aligned}\right\}(I)$$

In the neighborhood of each cluster point $\overline{\Theta}$, there are systems (iv) and (v) such that Θ_k is close to $\overline{\Theta}$ componentwise, μ_k is close to μ, and x^k is close to z. The solutions (x^k, μ_k) are unique and continuous functions of Θ_k. It follows, therefore, that $(x, \lambda) = (z, \mu)$, showing that (z, μ) must satisfy (I). Also z is unique. Since $\overline{\Theta}$ satisfies property (iii), it follows that z minimizes F on K. Thus, every cluster point of $\{x^k\}$ minimizes F on K, and $\{x^k\}$ converges to z.

EXERCISE

Prove z is unique.

C-2. *Nonpolyhedral Convex Programming II*

Our present techniques for minimizing convex functions[2] apply when the function has a continuous gradient or is polyhedral. We now consider the general situation when f is convex. Similarly, the constraint set will only be assumed to be closed and convex.

We digress temporarily to obtain some information about convex functions.

---

[2] Except in the special situation of Section C-1.

Lemma

Let S be a nonempty convex subset of E_n and f a convex function on S. Then

(i) If S' is a subset of S containing r points and $x \in H(S')$ with representation $x = \sum_{i=1}^{r} \lambda_i x^i$, then

$$f(x) \le \sum_{i=1}^{r} \lambda_i f(x^i).$$

(ii) If S is open and $y \in S$, there exists a neighborhood $N(y)$ and a number α' such that $f(x) \le \alpha'$ for all $x \in N(y)$.

Proof

(a). If $r = 2$, the lemma is true by definition. Assume, then, that the lemma is true when S' contains $r - 1$ points, and that $\lambda_1 \ne 0$. We may then write $x = \lambda_1 x^1 + (1 - \lambda_1)z$, where

$$z \sum_{i=2}^{r} \lambda_i = \sum_{i=2}^{r} \lambda_i x^i.$$

Thus, $f(x) \le \lambda_1 f(x^1) + (1 - \lambda_1)f(z)$. Since z is in the convex hull of $r - 1$, points out of S,

$$f(x) \le \sum_{i=1}^{r} \lambda_i f(x^i).$$

To prove (b), we observe that, by a translation if necessary, we may assume that $0 \in S$. Let C be a cube contained in S with radius α centered at the origin. Let $C' = \{x \in E_n : 0 < x_i < \alpha/n, 1 \le i \le n\}$. Thus, if $x \in C'$, and δ^i, $(1 \le i \le n)$, denotes the rows of the identity matrix,

$$x = \sum_{i=1}^{n} x_i \delta^i = \sum_{i=1}^{n} \frac{x_i}{\alpha} \alpha \delta^i + \left(1 - \sum_{i=1}^{n} \frac{x_i}{\alpha}\right) \cdot 0.$$

Therefore, x belongs to $H(\{\alpha \delta^i : 1 \le i \le n\} \cup \{0\})$ and

$$f(x) \le \frac{1}{n} \sum_{i=1}^{n} |f(\alpha \delta^i)| + |f(0)|.$$

Thus, if $x \in C'$, $f(x)$ is majorized.

Now take y arbitrarily in S and x^0 arbitrarily in C'. There exists, then, $n + 1$ points $x^1, \cdots, x^{n+1}$ in C', containing x^0 in the interiors of their convex hull. For example, we might take sufficiently small negative multiples of the rows of the identity together with a small positive multiple of the point $(1, \cdots, 1)$ for the points $x^i - x^0$, $1 \le i \le n + 1$.

Since S is open, for some $\gamma > 1$, the point $z - x^0 = \gamma(y - x^0) \in S$. Thus,

$$y - x^0 = \frac{1}{\gamma}(z - x^0) + \left(1 - \frac{1}{\gamma}\right)\left[\sum_{i=1}^{n+1} \lambda_i x^i - x^0\right],$$

with

$$\lambda_i > 0, (1 \le i \le n + 1)$$

whence

$$y = \sum_{i=1}^{n+1} \mu_i x^i + \mu_{n+2} z \quad \text{with} \quad \mu_i > 0, \qquad (1 \le i \le n + 2)$$

and

$$\sum_{i=1}^{n+2} \mu_i = 1.$$

For any vector ε (see Section B-2, p. 71) there exists σ_i, $(1 \le i \le n + 1)$ such that

$$\sum_{i=1}^{n+1} \sigma_i = 0 \quad \text{and} \quad \sum_{i=1}^{n+1} \sigma_i(x^i - x^0) = \sum_{i=1}^{n+1} \sigma_i x^i = \varepsilon.$$

For ε sufficiently small, $(\mu_i + \sigma_i) > 0$. Thus, there exists a neighborhood $N(y)$ such that $y' \in N(y)$ implies that

$$y' = \sum_{i=1}^{n+1} \mu'_i x^i + \mu'_{n+2} z,$$

with $\mu'_i > 0$, $(1 \le i \le n + 2)$, and

$$\sum_{i=1}^{n+2} \mu'_i = 1.$$

Finally, $f(y') \le f(z) + \bar{\alpha}$ for all $y' \in N(y)$ and some suitable number $\bar{\alpha}$.

Theorem

If f is a convex function defined on an open nonempty convex subset of E_n, f is continuous.

Proof

By the above lemma, if $x^0 \in S$, there is a sphere $N(x^0)$ of radius γ and number α such that if $x \in N(x^0)$, $f(x) \le \alpha$. Assume that $f(x^0) = 0$ and $x^0 = 0$.

Given $\varepsilon > 0$, choose $\delta < \varepsilon \gamma$. Then $\|x\| < \delta$ implies that x/ε and $-(x/\varepsilon)$ are in $N(x^0)$. Compute

$$f(x) = f\left((1 - \varepsilon) \cdot 0 + \varepsilon \frac{x}{\varepsilon}\right) \leq \varepsilon f\left(\frac{x}{\varepsilon}\right) \leq \varepsilon \alpha.$$

Also,

$$0 = f(0) = f\left(\left(-\frac{x}{\varepsilon}\right)\left(1 - \frac{1}{1 + \varepsilon}\right) + \left(\frac{1}{1 + \varepsilon}\right)x\right) \leq \frac{\varepsilon}{1 + \varepsilon} f\left(-\frac{x}{\varepsilon}\right) + \left(\frac{1}{1 + \varepsilon}\right) f(x).$$

Thus,

$$0 \leq \varepsilon f\left(-\frac{x}{\varepsilon}\right) + f(x),$$

showing that $f(x) \geq -\varepsilon \alpha$, and therefore f is continuous at x^0.

EXERCISES

1. Using Cramer's rule, show that in the above lemma there exists a number M such that $\|\sigma\| \leq M \|\varepsilon\|$. Using the inequality $\|\cdot\|_2 \leq \sqrt{n} \|\cdot\|_\infty$, estimate the radius of $N(y)$.

2. The set $S = \{(x, x_{n+1}) \in E_{n+1} : x_{n+1} \geq f(x)\}$ will be called the *supergraph* of f. Show that if f is u.s.c. (upper semicontinuous), then S has interior, and that if f is l.s.c., S is closed.

3. Prove that if C is closed and convex, then C is the intersection of all its supporting half-spaces (that is, the half-spaces that correspond to the supporting hyperplanes for C).

Hint: If x belongs to the intersection of the half-spaces and is not in C, look at the closest point in C to x.

Theorem

A convex function f defined on an open nonempty convex subset C of E_n is the supremum of a family of affine functions.

Proof

Let S denote the supergraph of f. Since f is continuous, S is closed, convex, and has interior. Thus, S is equal to the intersection of its supporting half-spaces. The boundary points of S are $\{(x, f(x)) : x \in C\}$, so that the supporting half-spaces can be indexed out of C. Namely, if $\bar{x} = (x, x_{n+1})$, then the supporting half-space to S at $(\xi, f(\xi))$ may be written

$$H_\xi = \{\bar{x} \in E_{n+1} : [\bar{\alpha}(\xi), \bar{x}] \leq b(\xi)\}.$$

Here $\bar{a}(\xi) \in E_{n+1}$ and $b(\xi) \in E_1$. In terms of inner products in E_n, we may write

$$H_\xi = \{\bar{x}: [\alpha(\xi), x] + \alpha_{n+1}(\xi)x_{n+1} \le b(\xi)\}.$$

We show that if $C^- = \{\xi \in C: \alpha_{n+1}(\xi) < 0\}$, then

$$S = \bigcup_{\xi \in C^-} H_\xi.$$

First of all, $\alpha_{n+1}(\xi) \le 0$, for otherwise, with x fixed and large x_{n+1}, $\bar{x} \in S$, but $\bar{x} \notin H_\xi$, contradicting that H_ξ is a supporting half-space for S. Take $\bar{x} \in S$; then $\bar{x} \in H_\xi$ for all $\xi \in C^-$. Conversely, assume $\bar{x} \in H_\xi$ for all $\xi \in C^-$ and $\bar{x} \notin S$. Since

$$S = \bigcap_{x \in C} H_\xi \quad \text{and} \quad \alpha_{n+1}(\xi) \le 0,$$

$\bar{x} \notin H_\xi$ for some ξ such that $\alpha_{n+1}(\xi) = 0$. For this ξ we have $[\alpha(\xi), x] > b(\xi)$. On the other hand, the point $(x, f(x)) \in S$; hence, $[\alpha(\xi), x] + \alpha_{n+1}(\xi)f(x) = [\alpha(\xi), x] \le b(\xi)$. This contradiction shows that

$$\bar{x} \in S \Leftrightarrow [\alpha_n(\xi), x] + \alpha_{n+1}(\xi)x_{n+1} \le b(\xi) \qquad \text{for all } \xi \in C^-.$$

Thus

$$\bar{x} \in S \Leftrightarrow x_{n+1} \ge \sup_{\xi \in C^-} \alpha_{n+1}^{-1}(\xi)\{b(\xi) - [\alpha_n(\xi), x]\},$$

from which we conclude that

$$f(x) = \sup_{\xi \in C^-} \{[\alpha'_n(\xi), x] - b'(\xi)]\},$$

where

$$\alpha'_n(\xi) = \frac{-\alpha_n(\xi)}{\alpha_{n+1}(\xi)} \quad \text{and} \quad b'(\xi) = \frac{-b(\xi)}{\alpha_{n+1}(\xi)}.$$

Remark

If the gradient of f is defined on C, then by Section D-3, proof of (c), $f(x) \ge f(\xi) + [\nabla f(\xi), x - \xi]$. Thus,

$$\alpha'_{n+1}(\xi) = \nabla f(\xi), \quad b'(\xi) = -([\nabla f(\xi), \xi] - f(\xi)),$$

and

$$f(x) = \max_{\xi \in C} \{[\alpha'_n(\xi), x] - b'(\xi)]\},$$

since the max is achieved at $\xi = x$.

We now turn our attention to the problem of minimizing a convex function f on a nonempty convex subset K of E_n. In order to render this problem tractable, we assume that the following data are available:

(i)   A family $\mathscr{F}_1$ of affine functions whose supremum is the function f.
(ii) A family $\mathscr{F}_2$ of affine functions such that
$$K = \{x\colon F_2(x) \leq 0 \qquad \text{for all} \qquad F_2 \in \mathscr{F}_2\}.$$
(iii) If $f_i(x) = \sup\{F_i(x)\colon F_i \in \mathscr{F}_i\}$, $(i = 1, 2)$, then, given $\varepsilon > 0$, a function $F'_i \in \mathscr{F}_i$ can be found with the property that $f_i(x) - F'_i(x) < \varepsilon$.

It is convenient to embody (i) and (ii) as follows: Let $\mathscr{A}$ and $\mathscr{B}$ denote bounded subsets of E_n, and a and b denote real-valued functions on $\mathscr{A}$ and $\mathscr{B}$, respectively. Set $K = \{x \in E_n\colon [B, x] - b(B) \leq 0 \;\forall\; B \in \mathscr{B}\}$ and
$$f = \sup\{[A, x] - a(A)\colon A \in \mathscr{A}\}.$$

Assume that K is not empty. Our final hypothesis is the following:

(iv) There exists finite subsets $\mathscr{A}^0 \subset \mathscr{A}$ and $\mathscr{B}^0 \subset \mathscr{B}$ respectively, with the property that for some number α, the set
$$S^0(\alpha) = \{x \in E_n\colon [A, x] - b(A) \leq \alpha \;\forall\; A \in \mathscr{A}^0$$
and $[B, x] - b(B) \leq 0 \;\forall\; B \in \mathscr{B}^0\}$ is nonempty and bounded.

Algorithm

At the mth step of the algorithm, two finite subsets $\mathscr{A}^m$ and $\mathscr{B}^m$ of $\mathscr{A}$ and $\mathscr{B}$, respectively, are given. Let
$$f^m = \max\{[A, x] - a(A)\colon A \in \mathscr{A}^m\} \quad \text{and} \quad K^m$$
$$= \{x\colon [B, x] \leq b(B) \;\forall\; B \in \mathscr{B}^m\}.$$
Select x^m to minimize f^m on K^m. Select $A' \in \mathscr{A}$ to maximize $[A, x^m] - b(A)$ and $B' \in \beta$ to maximize $[B, x^m] - b(B)$, respectively $[A, x^m] - b(B)$, respectively, within a tolerance of $1/m$. Begin anew with
$$\mathscr{A}^{m+1} = \mathscr{A}^m \cup \{A'\} \text{ and } \mathscr{B}^{m+1} = \mathscr{B}^m \cup \{B'\}.$$

Theorem

The algorithm is effective in the sense that

(i) $f^m(x^m) \uparrow \inf\{f(x)\colon x \in K\} = p$.
(ii) The sequence $\{x^m\}$ has cluster points, each of which lies in K and minimizes f thereon. If f has a unique minimum on K, $\{x^m\}$ converges.
(iii) The bounds $f^m(x^m) \leq p \leq f(x^m)$ hold and provide a control for terminating computation.

Proof

Let $S^m(\alpha) = \{x \in E_n : [A, x] - b(A) \le \alpha \not\vdash A \in \mathscr{A}^m\}$ and $[B, x] - b(B) \le 0 \not\vdash B \in \mathscr{B}^m\}$. Let

$$S(\alpha) = \{x \in K : [A, x] - b(A) \le \alpha \not\vdash A \in \mathscr{A}\}.$$

If $r \ge m$, $S^0(\alpha) \supset S^m(\alpha) \supset S^r(\alpha) \supset S(\alpha)$. By hypothesis, $S^0(\alpha)$ is bounded and nonempty for some α. By Section A-1, Problem 5, $S^0(\alpha)$ is bounded for all α. If $\alpha \ge \inf\{f(x) : x \in K\} = p$, then $S(\alpha)$ is nonempty. Since f is continuous, it has a minimum on $S(\alpha)$. Thus, the set $S^m(p)$ is compact and nonempty for $m = 0, 1, 2, \cdots$. Since

$$f^m(x^m) = \inf\{f^m(x) : x \in K^m\} \le \inf\{f^{m+1}(x) : x \in K^{m+1}\} = f^{m+1}(x^{m+1}) \le p,$$

it follows that $x^m \in S^0(p)$ for all m. Let y be any cluster point of $\{x^m\}$. Evidently,

$$y \in \bigcap_{m=0}^{\infty} S^m(p);$$

indeed $y \in S(p)$, for otherwise the numbers $g(y) = \sup\{[B, y] - b(B) : B \in \mathscr{B}\}$ or $f(y) - p$ are positive. We shall show that this is impossible.

Let $3\delta = f(y) - p$ and choose $m \ge 1/\delta$ so that $\|x^m - y\| < \delta/r$, where $r = \sup\{\|A\| : A \in \mathscr{A}\}$. Then

$$f(y) = \sup\{[A, x^m] + [A, y - x^m] - a(A)]\}$$
$$\le f(x^m) + \delta < [A', x^m] - a(A') + \frac{1}{m} + \delta$$
$$\le [A', x^m] - a(A') + 2\delta < [A', y] - a(A') + 3\delta$$
$$\le 3\delta + p$$

because $y \in S^{m+1}(p)$. Therefore, $f(y) < f(y)$, a contradiction. The same proof, *mutatis mutandis*, shows that $g(y) \le 0$. It follows, therefore, that y minimizes f on K.   Q.E.D.

Example

Let D be a compact subset of Euclidean r-space, E^r, and let $(f^0, \cdots, f^n)$ be continuous and linearly independent functions on D. Let

$$F(x) = \max\left\{\left|\sum_{i=1}^{n} x_i f^i(t) - f^0(t)\right| : t \in D\right\}.$$

Consider the problem of minimizing F subject to $x_i \ge 0, i = 1, \cdots, n$. We shall assume that for each x, a point t can be found so that

$$F(x) - |\sum x_i f^i(t) - f^0(t)| < \varepsilon.$$

Stated otherwise, we are assuming that points t can be found for which

$$\sum_{i=1}^{n} x_i f^i(t) - f^0(t)$$

is arbitrarily close to its maximum and minimum on D.

For each $t \in D$, let a subset $\alpha(t)$ of E_n be prescribed and set

$$\mathscr{A} = \bigcap_{t \in D} \alpha(t).$$

Specifically, let

$$A(t) = (f^1(t), \cdots, f^n(t)), \qquad \alpha(t) = \{A(t), -A(t)\}$$

and set

$$a(A) = f^0(t) \qquad \text{if} \quad A = A(t),$$
$$= -f^0(t) \qquad \text{if} \quad A = -A(t).$$

Then

$$\max\left\{ \left| \sum_{i=1}^{n} x_i f^i(t) - f^0(t) \right| : t \in D \right\} = \max\{[A, x] - a(A): A \in \mathscr{A}\}.$$

Let δ^i denote the rows of the $n \times n$ identity matrix and set $B^i = -\delta^i$. We have, then, that $K = \{[B^i, x] \le 0, 1 \le i \le n\}$. It remains to verify the hypothesis of (iv). The hypotheses that the set $\{f^i: 1 \le i \le n\}$ is linearly independent may be equivalently stated, as the set $\{(f^1(t), \cdots, f^n(t)): t \in D\}$ has 0 for its orthogonal complement in E_n. There exists, therefore, a subset of n points of D, say $\{t_1, \cdots, t_n\}$ such that the matrix with components $\{f^i(t_j): 1 \le i \le n, 1 \le j \le n\}$ has rank n. Arguing as in the proof of Exercise 4, of A-1, p. 56, we see that the set

$$S^0(\alpha) = \left\{ x \in E_n: \left| \sum_{i=1}^{n} x_i f^i(t_j) - f^0(t_j) \right| \le \alpha, \text{ and } x_j \ge 0, 1 \le j \le n \right\}$$

is bounded.

CHAPTER III

INFINITE DIMENSIONAL PROBLEMS

SECTION A. LINEAR SPACES AND CONVEX SETS

A-1. *Linear Spaces*

A *real linear space* is a set E and the field of reals R together with an operation $+: E \times E \to E$ and an operation $\cdot$ such that $R \times E \to E$ with the following properties: If x, y, z are in E and α, β are in R, then

   (i) $x + y = y + x$.

  (ii) $x + (y + z) = (x + y) + z$.

 (iii)   There exists θ in E such that $0 \cdot x = \theta$ for all x in E.

 (iv) $\alpha(\beta x) = (\alpha\beta)x$.

  (v) $(\alpha + \beta)x = \alpha x + \beta x$.

 (vi) $1 \cdot x = x$.

(vii) $\alpha(x + y) = \alpha x + \alpha y$.

The symbol 0 will generally be used ambiguously for θ and for the natural number 0.

A subspace E' of a real linear space E is a subset of E such that if x and y are in E', then $\alpha x + \beta y$ are in E'. An *affine subspace* E', or *linear variety* or *linear manifold*, in E is a subset such that if $x, y \in E'$ and $\alpha \in R$, then $\alpha x + (1 - \alpha)y$ is in E'. Observe that if A is an affine subspace, the set $S = \{x - z: x \in A\}$, with z fixed in A, is a subspace. The proof of this fact is an immediate consequence of the definitions. First we observe that if $x - z$ is in S, then so is $\lambda(x - z)$ for arbitrary λ because $\lambda(x - z) = \lambda x + (1 - \lambda)z - z$, with x and z in A. Assume that $x - z$ and $y - z$ belong to S. Then $\lambda(x - z) + \mu(y - z)$ belongs to S, for clearly $(\lambda x + \mu x)/(\lambda + \mu) - z$ is in S and therefore so is $\lambda x + \mu y - z(\lambda + \mu) = \lambda(x - z) + \mu(y - z)$.   Q.E.D.

Observe that if $z = \theta = 0$, then A is also a linear subspace and that the intersection of affine subspaces is an affine subspace. Let B be a subset of E. B is contained in some affine subspace because E itself is an affine subspace. The *least affine subspace* containing B is the intersection of all affine subspaces containing B and is called the *affine hull* of B. We similarly define the *linear hull* of B to be the intersection of all linear subspaces containing B. A set S in E is called *convex* if x and y in S imply $\lambda x + (1 - \lambda)y \in S$ for all λ such that $0 \le \lambda \le 1$. Thus, in the plane, a circle and a triangle are convex, but an annulus if not. The *convex hull* of a set S denoted by $H(S)$ is the intersection of all convex sets containing S. If S and T are subsets of E, the set $S + T = \{z = x + y: x \in S \text{ and } y \in T\}$. If α is a number, the set $\alpha S = \{\alpha x: x \in S\}$.

EXERCISES

1. Consider the linear space $C[0, 1]$ of all continuous functions on $[0, 1]$. Give examples of proper subsets of $C[0, 1]$ that are a subspace, a linear variety that is not a subspace, and a convex set.

2. Show that the linear hull of a set S is identical with the set of all finite linear combinations of S.

3. Show that the convex hull of a set S is the set of all finite sums of the form $\sum \lambda_i x_i$ with $x_i \in S \lambda_i \ge 0$ and $\sum \lambda_i = 1$.

A-2.   Normed Linear Spaces

A *real normed linear* space (NLS) E is a real linear space with a function $\|\cdot\|$ taking E into R such that

(i) $\|x\| \ge 0$.

(ii) $\|x\| = 0$ if and only if $x = 0$.

(iii) $\|\alpha x\| = |\alpha| \|x\|$.

(iv) $\|x + y\| \leq \|x\| + \|y\|$.

In all that follows the adjective "real" will be suppressed.

A normed linear space is a metric space under the metric $d(x, y) = \|x - y\|$. An NLS E is *complete* if every Cauchy sequence in E has its limit in E. Stated otherwise, every Cauchy sequence is convergent. A complete NLS is called *a Banach space*, or *B-space*.

E X E R C I S E

Prove that in a B-space every absolutely converging series converges, i.e.,

$$\sum_{k=1}^{\infty} \|x_k\| < \infty \Rightarrow \left\{ \sum_{k=1}^{K} x_k \right\} \to \sum_{k=1}^{\infty} x_k .$$

We now give some concrete examples of abstract B-space:

(a) In Chapter I, Sections B-1 and B-3, Exercise 2, we have seen that $C[0, 1]$ is a complete metric space. Hence $C[0, 1]$ is a B-space.

(b) For our next example we consider the space l_p of which the elements are sequences of numbers $(x_1, x_2, \cdots) = x$ such that

$$\left[\sum_{i=1}^{\infty} |x_i|^p \right]^{1/p} = \|x\|_p < \infty.$$

We now return to Hölder and Minkowski's inequalities and restate them for the space l_p. Let x and y be elements of the space l_p and l_q, respectively. Then

$$\sum_{i=1}^{\infty} |x_i y_i| \leq \|x\|_p \|y\|_q .$$

If $z \in l_p$, then $\|x + y\|_p \leq \|x\|_p + \|z\|_p$. To prove Hölder's inequality, recall Section D-7, Chapter I, and write

$$\lim_{n \to \infty} \sum_{i=1}^{n} |x_i y_i| \leq \lim_{n \to \infty} \left[\sum_{i=1}^{n} |x_i|^p \right]^{1/p} \left[\sum_{i=1}^{n} |y_i|^q \right]^{1/q} = \|x\|_p \|y\|_q .$$

Thus

$$\lim_{n \to \infty} \sum_{i=1}^{n} |x_i y_i| = \sum_{i=1}^{\infty} |x_i y_i| \leq \|x\|_p \|y\|_q .$$

Minkowski's inequality is proved in the same way. It follows that l_p is a normed linear space for $1 < p < \infty$. When $p = 1$ and $p = \infty$, it is easy to see that l_p is an NLS.

We now prove the completeness of l_p. Let $\{x^k\}$ be a Cauchy sequence in l_p. Clearly, $\|x^k - x^j\| \geq |x_i^k - x_i^j|$ for all i. Thus, for fixed i, the sequence $\{x_i^k\}$ converges to, say, x_i. Let $x = (x_1, x_2, \cdots)$. We show that $x \in l_p$ and $\{x^k\} \to x$. Given $\varepsilon > 0$, choose N so that $k, j \geq N$ implies $\|x^k - x^j\| < \varepsilon$. Then

$$\left[\sum_{i=1}^{n} |x_i^k - x_i^j|^p \right]^{1/p} < \varepsilon, \qquad \text{for } n = 1, 2, 3, \cdots.$$

Taking the limit as $k \to \infty$, we get

$$\left[\sum_{i=1}^{n} |x_i - x_i^j|^p \right]^{1/p} \leq \varepsilon,$$

whence $\|x - x^j\|_p \leq \varepsilon$ so that $\{x^k\} \to x$; and, moreover, $x - x^j \in l_p$. Thus, $x = x^j + x - x^j$. Since x^j belongs to l_p, so does x.   Q.E.D.

EXERCISES

1. Let M denote the space of all bounded functions on $[0, 1]$. For $f \in M$, $\|f\| = \sup\{|f(x)| : x \in [0, 1]\}$. Sketch a proof that M is a B-space. (*Hint:* Look over the proof that $C[0, 1]$ is a B-space.)

2. Let m denote the space of all bounded sequences. Prove that m is a B-space.

3. Let c denote the space of all converging sequences. Prove that c is a closed subspace of m and is thus a B-space.

There are other important Banach spaces, which we shall not discuss. As one example, there is the analog to the space l_p, which is the space $L_p[0, 1]$ of Lebesgue[1] measurable functions on $[0, 1]^2$ whose pth power is integrable. That is, we consider functions f for which

$$\int_0^1 |f(t)|^p \, dt < \infty.$$

The norm in $L_p[0, 1]$ is defined by

$$\|f\| = \left[\int_0^1 |f(t)|^p \, dt \right]^{1/p}$$

---

[1] The reader who is not familiar with Lebesgue integration is referred to W. Rudin, *Principles of Mathematical Analysis*, McGraw Hill, New York, 1957, p. 227, for information. This information, however, is not needed for the sequel.

[2] Lebesgue measure and $[0, 1]$ can also be replaced by arbitrary measures and measurable sets, respectively.

4. A B-space is *separable* if it is a separable metric space. Prove that the space l_p is separable. Prove that the space $C[0, 1]$ is separable. *Hint:* Consider points of the form $(x_1 \cdots, x_n, 0, \cdots)$ for $n = 1, 2, 3, \cdots$ and x_i rational. Recall Weierstrass' theorem for approximation continuous functions by polynomials.

5. Show that the convex hull of a compact subset U of l_2 is not necessarily closed. *Hint:* Let

$$U = \{(0, \cdots, 0), (1, 0, 0 \cdots), (0, \tfrac{1}{2}, \cdots), (0, 0, \tfrac{1}{3}, 0, \cdots) \cdots\}$$

$$= \{u^i : i = 1, 2, \cdots\}.$$

Let

$$x^k = \sum_{i=1}^{k} \lambda_i^k u^i,$$

where

$$\{\lambda_i^k\} = \{\tfrac{1}{2}, \tfrac{1}{4}, \tfrac{1}{8}, \ldots, 1/2^{k-1}, 1/2^{k-1}\}.$$

Show that $\lim\{x^k\}$ exists and is not in $H(U)$.

A-3. Hilbert Space

A (*real*) *Hilbert space* is a B-space that has a mapping $[\cdot, \cdot]$, called an *inner product*, taking $H \times H$ into R with the following properties:

(i) $[x, y] = [y, x]$.
(ii) $[\alpha x + \beta y, z] = \alpha[x, z] + \beta[y, z]$.
(iii) $[x, x] = \|x\|^2$.

EXERCISE

Show that in a Hilbert space the following hold:
1. $[x, 0] = [0, y] = 0$.
2. $|[x, y]| \leq \|x\| \|y\|$ (Schwarz inequality).
(*Hint:* $0 \leq [x - \lambda y, x - \lambda y]$ is a quadratic in λ with a nonpositive discriminant.) Show also that $[x, y] = \|x\| \|y\|$ only if $x = \alpha y$.
3. $\|x + y\|^2 + \|x - y\|^2 = 2\|x\|^2 + 2\|y\|^2$ (parallelogram law).
4. l_2 is a Hilbert space where $[x, y] = \sum x_i y_i$. (See Section A-2.)
The space $L_2[0, 1]$ (see Section A-2) is also a Hilbert space with

$$[f, g] = \int_0^l f(t)g(t)\,dt \qquad \text{for } f \text{ and } g \text{ in } L_2[0, 1].$$

A linear space that is equipped with an inner product is called an *inner product*, or *pre-Hilbert space*. Thus, a complete inner product space is a Hilbert space.

A-4. Convex Sets in Hilbert Space

In this section we shall consider some special properties of convex sets in Hilbert space and inner product spaces.

Theorem 1

If H is a Hilbert space, C a closed convex subset of H, and x a point of H, then there exists a unique point y in C closest to x; i.e.,

$$\|x - y\| = \inf\{\|x - z\| : z \in C\}.$$

Proof

Let $d = \inf\{\|x - z\| : z \in C\}$ and take a sequence a points $\{y_k\}$ in C such that $\{\|y_k - x\|\}$ converges downward to d. In the parallelogram law (Section A-3, Exercise 3), $\|a + b\|^2 + \|a - b\|^2 = 2\|a\|^2 + 2\|b\|^2$, set $y_n - x = a$ and $y_m - x = b$. Then

$$\|y_n - y_m\|^2 = 2\|y_n - x\|^2 + 2\|y_m - x\|^2 - \|y_n + y_m - 2x\|^2$$

$$= 2\|y_n - x\|^2 + 2\|y_m - x\|^2 - 4\left\|\frac{y_n + y_m}{2} - x\right\|^2$$

$$\leq 2\|y_n - x\|^2 + 2\|y_m - x\|^2 - 4d^2,$$

since $(y_n + y_m)/2$ is a point of C.

Since $\|y_m - x\|$ converges to d, the above inequality shows that $\{y_n\}$ is a Cauchy sequence. Because H is complete and C is closed, y belongs to C. It remains to prove that y is unique. If $\|y - x\| = \|y' - x\| = d$, then, as above, $\|y - y'\|^2 \leq 0$. Q.E.D.

We might observe that the element y above solves an "extremal" problem; namely, y minimizes the function $\|\cdot - x\|$ on the set C. We might ask at this juncture whether we can characterize the solution y of this extremal problem, since we know this is possible for other extremal problems. For example, the vanishing of a derivative is a necessary condition for a function to have an extremal on the interior of its domain. And with certain constrained extremal problems, we have the method of Lagrange multipliers, which we have applied to the extremal problem of determining the eigenvalues of a symmetric matrix in Chapter I, Section C-4. For the above problem, we have the following result.

Theorem 2

Let C be a convex set in an inner product space H. A point b in C is closest to a point a not in C if and only if $\beta = [x - b, b - a] \geq 0$ for all $x \in C$.

Proof

Assume that b is nearest a and take x arbitrarily in C. By the convexity of C, $tx + (1 - t)b \in C$ for all t, $0 \leq t \leq 1$. Thus, $0 \leq \|a - (tx + (1 - t)b\|^2 - \|a - b\|^2$, since b is nearest a. This inequality may be written as $t^2\|b - x\|^2 + 2t\beta \geq 0$. If $\beta < 0$, then for small t we find $t^2\|b - x\|^2 + 2t\beta < 0$, a contradiction. For the converse, assume $\beta \geq 0$. Then

$$\|x - a\|^2 - \|a - b\|^2 = [x, x] - 2[a, x]$$
$$+ 2[a, b] - [b, b] = [x - b, x - b] + 2\beta \geq 0,$$

showing that $\|a - b\| \leq \|x - a\|$ for all x in C. Q.E.D.

An equivalent characterization to the above is that b in C is nearest a point a not in C if and only if b maximizes the linear function $[x, a - b]$ on C. (Observe that $\beta \geq 0$ if and only if $[x, a - b] \leq [b, a - b]$.) In the terminology of Chapter II, Section A-2, $a - b$ is normal to the supporting hyperplane for C at b. Thus, if we define half-spaces and hyperplanes analogously to Chapter II, Section A-1, a geometric interpretation can be given to the above theorem, namely: *if b is closest in C to a not in C, then there is a hyperplane of support to C at b and $a - b$ is the normal to this hyperplane.* Thus C lies on one side (the "nonpositive" side) of the hyperplane $A = \{x \in H: [x, a - b] = [b, a - b]\}$, while a lies on the "positive" side, i.e., $[a, a - b] > [b, a - b]$. The hyperplane A is sometimes called a *separating hyperplane* because it separates the points a and the set C. We may state this formally as a theorem.

Theorem 3

Let C be a closed convex subset of H and a a point not in C. There exists a hyperplane $A = \{x: [\xi, x] = \alpha\}$ separating C and a; i.e., $[\xi, x] \leq \alpha \leq [\xi, a]$ for all x in C.

Proof

Set $\xi = a - b$ and $\alpha = [b, a - b]$, where b in C is nearest to a.

A hyperplane $A = \{x: [\xi, x] = \alpha\}$ is said to *strictly separate* the point a from the convex set C if $[\xi, x] < \alpha < [\xi, a]$ for all x in C.

E X E R C I S E

Prove the following result: Let C be a closed convex subset of H and a a point not in C. There exists a hyperplane $A = \{x: [\xi, x] = \alpha\}$ strictly separating C and a. *Hint:* Observe that $[a - b, x - a] \le [a - b, b - a]$ for all x in C.

A-5. Projection Operator for C

We have seen that, given a closed convex set C and a point $x \in H$, there exists a unique point, say y, in C which is closest to x. We shall denote by P the mapping that assigns $y = Px$ to the point x. We call P the *projection operator* for C. We shall see below that the projection operator is continuous —in fact, Lipschitz continuous.

Theorem 1

The projection operator P for a closed convex set K in Hilbert space satisfies the Lipschitz condition $\|Px - Py\| \le \|x - y\|$, equality holding only if $x - Px = y - Py$.

Proof

Since Py is closest to y in K and Px belongs to K, $[Px - Py, Py - y] = \alpha \ge 0$; this follows from Theorem 2 of Section A-4, if $y \notin K$. If $y \in K$, $\alpha = 0$ because $y = Py$. Similarly, $[Py - Px, Px - x] = \beta \ge 0$. Thus, $\alpha + \beta \ge 0$ implies that $\|Px - Py\|^2 \le [Px - Py, x - y] \le \|Px - Py\|\|x - y\|$ by the Schwarz inequality, whence $\|Px - Py\| \le \|x - y\|$. Assume that $\|Px - Py\| = \|x - y\|$; then $[Py - Px, y - x] = \|Px - Py\|^2 = \|Px - Py\|\|x - y\|$. Hence, $Py - Px = y - x$.

Our attention will now be directed toward the following problem: Given two closed convex subsets of H, namely, K_1 and K_2, find points x' in K_1 and y' in K_2 such that $\|x' - y'\| \le \|x - y\|$ for all x in K_1 and y in K_2. We call $\|x' - y'\|$ *the distance from* K_1 *to* K_2. We shall see that the solution to this problem has a direct application to the solution of the problem of linear inequalities, which was discussed in Chapter II, Sections A-1 and B-2.

Lemma

Let K_1 and K_2 be two closed convex sets in Hilbert space. Let P_i denote the projection operator for K_i. Any fixed point of $Q = P_1 P_2$ is a point of K_1 nearest K_2, and conversely.

Proof

Assume that x is a fixed point of Q so that $x = P_1 P_2 x$. Set $y = P_2 x$. If $y = x$, the distance is 0 and we are done; otherwise, $x \notin K_2$ and $y \notin K_1$. Since x is nearest to y for all $u \in K_1$, we have that $[u, x - y] \geq [x, x - y]$. Similarly, for all $v \in K_2$ $[v, y - x] \geq [y, y - x]$, and adding, we get $[u - v, x - y] \geq \|x - y\|^2$, whence

$$\|u - v\| \geq \left[u - v, \frac{x - y}{\|x - y\|} \right] \geq \|x - y\|,$$

by Schwarz's inequality.

For the converse, suppose that x is a point of K_1 nearest to K_2. Then $\|x - P_2 x\| \leq \|u - v\|$ for all $u \in K_1$ and $v \in K_2$. Clearly, x is closest in K_1 to $P_2 x$ in K_2, whence $x = P_1 P_2 x$. Q.E.D.

Theorem 2

Let K_1 and K_2 be two closed convex sets in Hilbert space H and Q, the composition $P_1 P_2$ of their projection operators. Assume that x is arbitrary in K_1. Convergence of $\{Q^n x\}$ to a fixed point of Q is assured when either (a) one set is compact, or (b) one set is finite dimensional and the distance between the sets is attained.

Proof

The proof consists in showing that Q is a subcontractor with compact range. (See Chapter I, Section B-4.) We must show that

(a) $\|Qx - Qy\| \leq \|x - y\|$ if $(x, y) \in H \times H$.

(b) $x \neq Qx$ only if $\|Qx - Q^2 x\| < \|x - Qx\|$.

(c) $\{Q^n x\}$ has cluster points.

We have $\|P_1 P_2 x - P_1 P_2 y\| \leq \|P_2 x - P_2 y\| \leq \|x - y\|$; hence (a) is satisfied. To show that (b) is satisfied, assume $x \neq Qx = y$. Then $\|y - P_2 y\| \leq \|y - P_2 x\| = \|P_1 P_2 x - P_2 x\| < \|x - P_2 x\|$ because $P_1 P_2 x$ in K_1 is closest to $P_2 x$ in K_2, the strict inequality obtaining because $x \neq P_1 P_2 x$ and because Qx is the unique closest point to $P_2 x$. Since

$$\|y - P_2 y\| \neq \|x - P_2 x\|, \ \|Qx - Qy\| \leq \|P_2 x - P_2 y\| < \|x - y\|$$

by Theorem 1. Thus, $\|Qx - Q^2 x\| < \|x - Qx\|$.

The compactness of K_1 suffices for condition (c). If K_2 is compact, replace Q by $Q' = P_2 P_1$ and repeat the above argument. If y is a fixed point of Q', $P_1 y$ is a fixed point of Q. If K_1 is finite dimensional and the distance between K_1 and K_2 is achieved, then by the above lemma, Q has a fixed point, say y. We see that $\|Q^n x\| = \|Q^n x - Qy + y\| \le \|Q^n x - Q^n y\| + \|y\| \le \|x - y\| + \|y\|$. Thus, $\{Q^n x\}$ lies in a sphere, which is compact by the Bolzano-Weierstrass theorem.

The distance between two closed convex sets is not necessarily achieved; see Remark in Section A-6.

A-6. Distance Between Polytopes

Recall that a *polytope* P is a subset of E_n which is the intersection of a finite number of half-spaces. Thus, if P is a polytope, there exists points $A^i \in E_n$ and $b^i \in R$, $(1 \le i \le m)$, such that $P = \{x : [A^i, x] \le b^i, 1 \le i \le m\}$. A *face* of P is the set

$$F(i_1, i_2, \cdots, i_r) = \{x \in P : [A^{ij}, x] = b^{ij}, (1 \le j \le r), r \le m\}.$$

If dim P (dimension of P) is $r < n$, P lies on an affine subspace. We write $P = R \cap Q$, $R = \{x : [A^k, x] = b^k, (1 \le k \le r)\}$ and $Q = \{x : [A^i, x] \le b^i, (1 \le i \le q)\}$, with appropriate A^i and b^i. A *proper face* of P is the set

$$\{x \in R \cap Q : [A^{ij}, x] = b^{ij} (1 \le j \le q)\}.$$

Example

Let A_1 and A_2 be affine subspaces in E_n. The distance $d(A_1, A_2)$ between A_1 and A_2 is achieved.

Proof

Recall Section A-1, Chapter III, and observe that if A is an affine subspace, then $A = S + z$, where S is some subspace and z is a point of A. Let $A_1 = S_1 + z_1$ and $A_2 = S_2 + z_2$. Let $\{u_1, \cdots, u_r\}$ be a basis for S_1 and $\{v_1, \cdots, v_k\}$ be a basis for S_2. Thus, if $x \in A_1$ and $y \in A_2$, then

$$x = \sum \alpha_i u_i + z_1, \quad y = \sum \beta v_i + z_2,$$

and

$\inf\{\|x - z\| : x \in A_1 \text{ and } y \in A_2\}$

$$= \inf\{\left\| \sum \alpha_i u_i + z_1 - \sum \beta_i v_i - z_2 \right\| : \alpha \in E_r \text{ and } \beta \in E_k\}.$$

But $\inf\left\|\sum \alpha_i u_i - \sum \beta_i v_i - (z_2 - z_1)\right\|$ is the distance from the point $z = z_2 - z_1$ in E_n to the subspace S spanned by the linear hull of $\{u_1, \cdots, u_r, v_1, \cdots, v_k\}$. Since S is a closed convex subset of E_n, $d(z, S)$ is achieved. It follows that $d(A_1, A_2)$ is also achieved.   Q.E.D.

Theorem

Let P_1 and P_2 be polytopes in E_n. Then $d(P_1, P_2)$ is achieved.

Proof

We proceed by induction on the dimension of P_1. For the case dim $P_1 = 0$, P_1 is a point and P_2 a closed convex set. We must show that if the theorem is true when dim $P_1 \leq k$, $1 \leq k \leq n - 1$, it is true when dim $P_1 = k + 1$. The faces of P_1 are finite in number and are polytopes of dimension $k - 1$ or less; moreover the distance between each face and P_2 is achieved. Let d denote the minimum of these distances. If $d = d(P_1, P_2)$ we are done; otherwise there exists points $x^0 \in P_1$ and $y^0 \in P_2$ such that $d(x^0, y^0) < d$. If $k = n - 1$, then P_1 has interior and x^0 lies therein. Hence $[A^i, x^0] < b^i$, $1 \leq i \leq m$. For some point x' in the interior of the segment $\langle x^0, y^0 \rangle$, some index, (say i^0) satisfies $[A^{i_0}, x'] = b^{i_0}$, while $[A^i, x'] \leq b^i$, $1 \leq i \leq m$. But $d(x', y^0) < d(x^0, y^0)$, and x' lies on a face of P_1 which is a contradiction. Assume $k < n - 1$. If $d > d(P_1, P_2)$, choose z^0 (see Exercise below) to minimize the distance between P_2 and the affine hull of P_1. Then $\langle x^0, z^0 \rangle$ contains no point of a proper face of P_1, and therefore $z^0 \in P_1$.

EXERCISE

Prove that the distance from an affine subspace to a polytope is achieved.

Remarks

If either P_1 or P_2 is bounded, the proof is easily accomplished by a compactness argument. It is not true that the distance between two closed convex sets is achieved. Consider the following closed convex sets in E_2:

$$\left\{ (x, y): y \geq \frac{1}{x} \quad \text{and} \quad 1 \leq x \leq \infty \right\}$$

and

$$\{(x, y): y \leq 0 \quad \text{and} \quad 1 \leq x \leq \infty\}.$$

A-7.   On Linear Inequalities

An *orthant* is the set $0 = \{y \in E_m: y_i \leq b_i$, b fixed in $E_m\}$. Let A be an $m \times n$ matrix and b an m-vector. The system of inequalities $Ax \leq b$ is consistent if and only if the range of A meets 0. The range of A, being a subspace of

E_m, is a polytope. The orthant 0 is also a polytope. If we can determine the projection operators for these polytopes, then we can iterate their composition and find the distance d between these polytopes. If $d > 0$, the system of inequalities is inconsistent. If $d = 0$, it is consistent, and Theorem 2 may be applied to generate a sequence converging to the intersection of 0 and the range of A.

A little reflection will convince one that the value of the projection mapping for the orthant at z is $P_1 z = x$, where $x_i = z_i$ if $z_i \le b_i$ and $x_i = b_i$ if $z_i > b_i$. As a result of Exercise 1 below, if A is of rank n, $P_2 = A(A^*A)^{-1}A^*$. Furthermore, there is no loss of generality in assuming that A is of rank n.

EXERCISES

1. Let A be an $m \times n$ matrix and b an m-vector. Show that if x minimizes $\|Ax - b\|$, then $A^*(Ax - b) = 0$, and consequently if A has rank n, $x = (A^*A)^{-1}A^*b$.

Hints:

(a)   Prove first that if S is an affine subspace of an inner product space and $b \notin S$, then $[b - a, x - a] = 0$ for all $x \in S$ if and only if $a = P(b)$, where P is the projection operator for S.

(b)   Show that if S is the linear hull of $\{A^1, \cdots, A^n\}$,

$$[b - P(b), A^j] = 0, (1 \le j \le n).$$

Now take S to be the range of A and conclude that $[A^j, Ax - b] = 0$, $(1 \le j \le n)$, where A^j denotes the jth column of A.

(c) A^*A has rank n if A has rank n.

2.   Consider the case $m = 2, n = 1$. Give a picture where the system is consistent and where it is inconsistent. Prove that the above process of Theorem 2 converges as fast as a geometric progression for these examples.

Remark

If A has rank $r \le n$, the matrix A can be replaced by an $m \times r$ matrix A_0 whose r columns are the first r linearly independent columns of A. Then define $x_0 = (A_0^*A_0)^{-1}A_0^*b$ and $x = x'_0 + u$, where u is any vector in the null space of A and x'_0 is a n-tuple formed from the r-tuple x_0 by adding appropriate zero components. Then $A_0 x_0 = Ax$.

SECTION B. MISCELLANY

B-1. Linear Operators

Let E and F be normed linear spaces and A a map from E to F. For simplicity, the norms on E and F will be denoted by the same symbol $\|\cdot\|$, as will also the norm on linear operators defined below.

Definition

A linear operator A from E to F is a map satisfying $A(\alpha x + \beta y) = \alpha Ax + \beta Ay$ for all real α and β for all x and y in E.

Definition

The operator A is said to be continuous at $x \in E$ if, given $\varepsilon > 0$, there exists $\delta > 0$ such that $\|y - x\| < \delta$ implies that $\|Ay - Ax\| < \varepsilon$.

Definition

The operator A is said to be bounded if there exists a number $\beta > 0$ such that $\|Ax\| \le \beta \|x\|$ for all x in E. Define $\|A\| = \inf\{\beta : \|Ax\| \le \beta\|x\| : x \in E\}$.

Theorem 1

The following are equivalent:

(i) A is continuous at some point x in E.

(ii) A is continuous throughout E.

(iii) A is bounded on E.

(iv) $\sup\left\{ \left\| A\,\dfrac{x}{\|x\|} \right\| : x \ne 0 \right\} = \inf\{B : \|Ax\| \le B\|x\| : x \in E\} = \|A\|.$

Proof

(i) $\Rightarrow$ (ii). A is continuous at x if and only if $\{Ax_n\} \to Ax$ when $\{x_n\} \to x$. Let $\{y_n\} \to y$ and set $x_n = y_n - y + x$. Then $Ay_n = A(x_n - x) + Ay$. Thus, $\{Ay_n\} \to Ay$ and A is continuous at y.

(ii) $\Rightarrow$ (iii). If A is not bounded on E, for every $n = 1, 2, 3, \cdots$ there exists x_n such that $\|Ax_n\| > n\|x_n\|$. Set

$$y_n = \frac{1}{n}\frac{x_n}{\|x_n\|}.$$

Then $\|y_n\| = 1/n$, while

$$\|Ay_n\| = \frac{1}{n\|x_n\|}\|Ax_n\| > 1.$$

Consequently, A is not continuous at 0.

(iii) $\Rightarrow$ (iv).   The proof of this may be found in Chapter I, Section C-2.

(iv) $\Rightarrow$ (i).   By (iv), $\|Ax\| \leq \|A\|\|x\|$; hence, A is continuous at 0.

Theorem 2

Let E and F be normed linear spaces and let $B(E, F)$ denote the set of bounded linear operators from E to F. Assume that F is complete. Then $B(E, F)$ is a Banach space.

Proof

If we denote the zero operator in $B(E, F)$ by the operator that maps E into 0, we may verify that $B(E, F)$ is a normed linear space under the norm defined in Section B-1. We now show that this space is actually complete, even if E is not complete. Take a Cauchy sequence of operators $\{A_n\}$. Then $\|A_n - A_m\| < \varepsilon/2$ for $n, m > n(\varepsilon)$. Since F is complete, we have for each $x \in E$ that $\lim\{A_n x\}$ exists. This limit is a function of x, which we denote by Ax. We show that A is linear, $\lim\{A_n\} = A$ and A is bounded. Firstly,

$$A(\alpha x + \beta y) = \lim\{A_n(\alpha x + \beta y)\} = \lim\{[\alpha A_n x + \beta A_n y]\} = \alpha Ax + \beta Ay.$$

Secondly,

$$\|Ax - A_n x\| \leq \|Ax - A_m x\| + \|A_m - A_n\|\|x\|.$$

Letting $m \to \infty$, we see that for some N, $n > N$ implies that $\|A - A_n\| < \varepsilon/2$ and $\|A_n\| \leq \varepsilon/2 + \|A_N\|$. Finally, $\|Ax\| < (\varepsilon + \|A_N\|)\|x\|$, and therefore A is bounded.

Definition

If F is the space R of reals, the totality of bounded linear operators $B(E, R)$ is called *conjugate* space of the norm linear space E and is denoted by E^*. The elements of $B(E, R) = E^*$ are then called *bounded linear functionals*.

Remarks

(a)   We have proved that the conjugate $B(E, R)$ of a normed linear space E is Banach space.

Observe that E^* is equipped with the norm

$$\|f\| = \sup_{\|x\| = 1} |f(x)|,$$

and that the value of $\|f\|$ depends on the choice of norm in E.

(b)   Suppose E is a Hilbert space. Let $f(x) = [a, x]$, where a is fixed in E. Then f is a linear functional, which is bounded, as is evident by Schwarz's inequality. Thus, for every element in E we can define a linear functional in E^*.

(c)   A function f is said to "separate" distinct points x and y if $f(x) \neq f(y)$. There are linear functionals in the conjugate of a Hilbert space which separate points. For example: Let $f(x) = [b - a, x]$. Then $f(b) - f(a) = \|b - a\|^2 > 0$. (A like functional[3] that separates points in a Banach space will be obtained in Chapter III, Section C1, Corollary 2, p. 134. The importance of such a functional is in the following application. Suppose $f(x) = f(y)$ for all $f \in E^*$. Then we may conclude that $x = y$, for taking the particular functional $g = [x - y, \cdot]$, we have $0 = g(x) - g(y) = g(x - y) = \|x - y\|^2$. Hence, $x = y$.

We next give some examples of representing the conjugates of an NLS. In Section B-3, it will be shown that if H is a Hilbert space, then any element $f \in H^*$ can be identified with a unique element, say x_f, in H. By this is meant that $f(x) = [x_f, x]$ for all $x \in H$.

Consider now the space l_p. We shall show that every linear functional $f \in l_p^*$ can be identified with an element $y(f)$ of l_q, $(1/q + 1/p = 1)$ in the sense that $f(x) = [y(f), x]$ for all $x \in l_p$, where

$$[y(f), x] = \sum_{i=1}^{\infty} y_i(f)x_i.$$

A *topological basis* in an NLS E is a set of vectors $\{e^1, e^2, \cdots\}$ such that every $x \in E$ can be represented uniquely as

$$x = \sum_{i=1}^{\infty} \lambda_i e^i$$

where λ_i, $i = 1, 2, \cdots$ are scalars. This representation means that

$$\lim_{n \to \infty} \left\| x - \sum_{i=1}^{n} \lambda_i e^i \right\| = 0.$$

In the space l_p, $(p > 1)$, the vectors $(1, 0, 0 \cdots) = \delta^1$, $(0, 1, 0, 0 \cdots) = \delta^2$, $\cdots$ are a basis. Thus, for $x \in l_p$ and $f \in l_p^*$,

$$f(x) = f\left(\sum_{i=1}^{\infty} x_i e^i\right) = \sum_{i=1}^{\infty} x_i f(e^i).$$

---

[3] There exists a functional f, $\|f\| = 1$ such that $f(y - x) = \|y - x\|$.

Set $y_i(f) = f(e_i)$ and

$$x_i = \begin{cases} |y_i(f)|^{q-1} \operatorname{sgn} y_i(f) & \text{if } i \leq N \\ 0 & \text{if } i > N \end{cases}$$

Then

$$f(x) = \sum_{i=1}^{N} |y_i(f)|^q \leq \|f\| \|x\|$$

$$= \|f\| \left(\sum_{i=1}^{N} |y_i(f)|^{(q-1)p} \right)^{1/p}$$

$$= \|f\| \left(\sum_{i=1}^{N} |y_i(f)|^q \right)^{1/p},$$

whence

$$\left(\sum_{k=1}^{N} |y_i(f)| \right)^{1/q} \leq \|f\|.$$

Letting $N \to \infty$, $\|y(f)\|_q \leq \|f\|$. Thus, $y(f) \in l_q$.

Corollary (Banach-Steinhaus)

Let E be an NLS and F a B-space. Let $\{A_n\}$ be a sequence in $B(E, F)$. Assume that for all n, $\|A_n\| \leq M$, and $\lim\{A_n x\}$ exists for all x in any set that is dense in E. Then there exist $A \in B(E, F)$ such that $\{A_n x\} \to Ax$ for all $x \in E$.

Proof

Suppose that D is dense in E. Given an arbitrary $x \in E$, choose $x' \in D$ such that $\|x - x'\| < \varepsilon$. For sufficiently large m and n, we have $\|A_m x' - A_n x'\| < \varepsilon$, whence

$$\|A_m x - A_n x\| \leq \|A_m x' - A_n x'\| + \|A_m x - A_m x'\|$$

$$+ \|A_n x - A_n x'\| < \varepsilon + (\|A_m\| + \|A_n\|)\|x - x'\| < (2M + 1)\varepsilon.$$

Since F is complete $\{A_n x\} \to Ax$, and A is linear. Moreover

$$\|Ax\| = \lim_{n \to \infty} \|A_n x\| \leq \liminf \|A_n\| \, \|x\| \leq M\|x\|.$$

Remark

The converse of Theorem 2 is true, namely, if $\{A_n x\} \to Ax$ for all $x \in E$ then $\|A_n\| \leq M$. The proof is rather long. It will be omitted because it is not needed in the sequel.

B-2. *Application to Mechanical Quadrature*

In numerical analysis, definite integrals are calculated by a process called *mechanical quadrature*, or numerical integration. The process works as follows. Given $x \in C[0, 1]$, we wish the value of

$$f(x) = \int_a^b x(t)dt.$$

Let $\{t_1^n, \cdots, t_n^n\}$ be a subset of $[a, b]$ containing n distinct points. There is a unique polynomial P of degree $n - 1$ such that $x(t_k) = P(t_k)$, $1 \le k \le n$. A convenient expression for P is the Lagrange interpolation polynomial:

$$P(t) = \sum_{k=1}^n x(t_k^n)\lambda_k^n(t),$$

where

$$\lambda_i^n(t) = \frac{\pi(t)}{t - t_i^n}\,\pi'(t_i), \qquad \pi(t) = \prod_{j=1}^n (t - t_j^n)$$

and

$$\pi'(t_i) = \prod_{\substack{j=1 \\ j \ne i}}^n (t_i^n - t_j^n).$$

It is left to the reader to verify that, thus defined, $P_n(t)$ is a polynomial of degree $n - 1$, which assumes the values (*interpolates* x) $x(t_k^n)$ at $t = t_k^n$, $(1 \le k \le n)$. The idea of mechanical quadrature is to integrate this polynomial to obtain an approximation for $f(x)$.

Observe now that the formulas

$$f(x) = \int_a^b x(t)\,dt$$

and

$$f_n(x) = \sum_{k=0}^n x(t_k^n) \int_a^b \lambda_k^n(t)\,dt = \sum_{k=0}^n x(t_k^n)\mu_k^n$$

define points f, f_n, $n = 1, 2, 3, \cdots$ in the conjugate $C^*[a, b]$ of $C[a, b]$ because $\|f\| = b - a$ and

$$\|f_n\| = \sum_{k=1}^n |\mu_k^n|, \qquad (n = 1, 2, \cdots).$$

The sum for $f_n(x)$ is called a *quadrature formula*. Observe that μ_k^n is independent of x and can be calculated once and for all.

Theorem (Szego)

If

$$\sum_{k=1}^{n} |\mu_k^n| \le M \qquad (n = 1, 2, 3, \cdots),$$

then $\{f_n(x)\} \to f(x)$ for all $x \in C[a, b]$.

Proof

The set of all polynomials D is dense in $C[a, b]$. If x is a polynomial, say, of degree N and if $n \ge N + 1$,

$$x(t) = \sum_{k=1}^{n} x(t_k^n)\lambda_k^n(t).$$

Hence, $\{f_n(x)\} \to f(x)$ for all $x \in D$ and $\|f_n\| \le M, (n = 1, 2, 3, \cdots)$. An appeal to the Banach-Steinhaus theorem concludes the proof.

Remark

By virtue of the remark which follows the above Corollary, the condition

$$\sum_{k=1}^{n} |\mu_k^n| \le M$$

is also necessary.

Remark

If the coefficients μ_k^n are nonnegative, then for some M,

$$\sum_{k=0}^{n} |\mu_k^n| \le M \qquad (n = 1, 2, 3, \cdots).$$

Proof

If $x(t) \equiv 1$,

$$b - a = \int_a^b dt = \lim_{n \to \infty} \sum_{k=1}^{n} \mu_k^n.$$

Our object now is to show that there exists a selection of the points $\{t_k^n\}$ such that the numbers μ_k^n are positive. Specifically, we show the startling result that if, for each n, the numbers $\{t_1^n, \cdots, t_n^n\}$ lie at the roots of an nth degree member of any family of orthogonal polynomials on $[a, b]$, the resulting quadrature formula (discovered by Gauss) is exact if $x(t)$ is a polynomial of degree $\leq 2n - 1$. Moreover, the numbers μ_k^n are positive.

EXERCISES

1. Prove that if P is a polynomial satisfying $x(t_k^n) = P(t_k^n)$, $(1 \leq k \leq n)$, then P is unique. Assume P is of degree $n-1$.

2. Show that

$$\sum_{k=1}^{n} x(t_k^n) \lambda_k^n(t)$$

is a polynomial of degree $n - 1$, with values $x(t_k^n)$ at t_k^n, $(1 \leq k \leq n)$.

We now review the Gram-Schmidt process from linear algebra.

Lemma

Let I be an inner product space and let $\{x^i \in I : 1 \leq i \leq k\} = S_k$ be a linearly independent set in I. There exists, then, an orthogonal set $0_k = \{0 \neq y^i \in I : 1 \leq i \leq k\}$ (that is, $[y^i, y^j] = 0, i \neq j$), with the same linear span as S_k. Moreover,

$$y^1 = x^1 \qquad\qquad x^1 = y^1$$

$$y^2 = x^2 + \phi_{21} x^1 \qquad\qquad x^2 = y^2 + \theta_{21} y^1$$

$$y^3 = x^3 + \phi_{31} x^1 + \phi_{32} x^2 \qquad\qquad x^3 = y^3 + \theta_{31} y^1 + \theta_{32} y^2$$

$$\vdots \qquad\qquad\qquad\qquad \vdots$$

$$y^k = x^k + \phi_{k1} x^1 + \cdots \phi_{k,k-1} x^{k-1} \qquad x^k = y^k + \theta_{k1} y^1 + \cdots \theta_{k,k-1} y^{k-1}.$$

Proof

Assume that for $m \leq k - 1$, the above facts are true concerning the sets S_k and 0_k. Set

$$y^k = x^k - \theta_{k1} y^1 - \theta_{k2} y^2 - \cdots - \theta_{k,k-1} y^{k-1}.$$

If $\theta_{ki} = [x^k, y^i]/[y^i, y^i]$, then clearly $[y^k, y^i] = 0$, $(1 \leq i \leq k - 1)$. Replacing y^i, $(1 \leq i \leq k - 1)$ by its expression as a linear combination of $x^1, \cdots, x^i$, we get $y^k = x^k + \phi_{k1} x^1 + \cdots + \phi_{k,k-1} x^{k-1}$. If $y^k = 0$, we contradict the linear independence of the set $\{x^1, \cdots, x^k\}$. Clearly, z belongs to linear span of

$\{x^1, \cdots, x^k\}$ if and only if z belongs to the linear span of $\{y^1, \cdots, y^k\}$. Q.E.D.

The *Gram-Schmidt* process is defined as follows: Given a set $\{x^1, \cdots, x^m\}$, an orthogonal set $\{y^1, \cdots, y^m\}$ with identical linear span is given by

$$y^1 = x^1 \quad y^m = x^m - \sum_{i=1}^{m-1} \frac{[x^m, y^i]}{[y^i, y^i]} y^i, \quad m = 2, 3, \cdots, k.$$

The vectors $y^2, y^3, \cdots, y^k$ are determined successively.

We define an inner product space $I[a, b]$ by introducing an inner product in $C[a, b]$ defined by

$$[f, g] = \int_a^b f(t)g(t) \, dt.$$

Two functions f and g in $I[a, b]$ are said to be *orthogonal* if $[f, g] = 0$.

EXERCISES

1. Consider the space $I(-1, 1)$. Observe that the set $\{x^1, \cdots, x^n\}$, where $x^i = t \to t^{i-1}$, is linearly independent. Construct y^1, y^2, y^3. You should find that $y^1 = t \to 1$, $y^2 = t \to t$, and $y^3 = t \to t^2 - \frac{1}{3}$.

2. Prove that the polynomial y^m is orthogonal to any polynomial of degree less than $m - 1$.

The set $\{y^1, \cdots, y^n\}$ is often called a *family of orthogonal polynomials*. We now prove the following fact about orthogonal polynomials.

Lemma

Let y^n be a polynomial of degree $n - 1$ which is a member of an orthogonal family in $I(a, b)$. Then the $n - 1$ roots of y^n are all simple and lie in $[a, b]$.

Proof

Since y^n is orthogonal to y^1,

$$\int_a^b y^n(t) \, dt = 0.$$

It follows that y^n must have a root of odd multiplicity on $[a, b]$. Let t_i, $(1 \leq i \leq r < n - 1)$, denote the roots of y^n of odd multiplicity in $[a, b]$, and set

$$p(t) = \prod_{i=1}^{r} (t - t_i).$$

Then, by Exercise 2,

$$\int_a^b y''(t)p(t)\, dt = 0.$$

On the other hand, the polynomial $y''p$ has all roots of even multiplicity on $[a, b]$. This contradiction establishes the lemma.

We now return to the consideration of the coefficients μ^k for Gaussian quadrature. As we have seen:

$$\mu_k = \int_a^b \lambda_k(t)\, dt,$$

where

$$\lambda_i(t) = \frac{\pi(t)}{(t - t_i)\pi'(t_i)} \qquad \pi(t) = \prod_{j=1}^n (t - t_j)$$

and

$$\pi'(t_i) = \prod_{\substack{j=1 \\ j \neq i}}^n (t_i - t_j).$$

Recall that the quadrature formula

$$\int_a^b x(t)\, dt = \sum_{k=1}^n \mu_k x(t_k) \tag{I}$$

is exact if x is a polynomial of degree $\leq n - 1$ whenever $\{t_k : 1 \leq k \leq n\}$ is any subset of n distinct points of $[a, b]$. If, for some subset $\{t_k\}$ of n distinct points of $[a, b]$, the quadrature formula (I) is exact when x is any polynomial of degree $\leq 2n - 1$, this subset $\{t_k\}$ will be called a *Gauss set*.

Theorem

A necessary and sufficient condition for $\{t_k\}$ to be a Gauss set is that the polynomial π must have the same roots as the orthogonal polynomial y^{n+1}.

Proof

We prove first the necessity. Let p be any polynomial of degree $\leq n - 1$. The product $f = \pi p$ is of degree $\leq 2n - 1$, whence

$$\int_a^b f(t)\, dt = \sum_{k=1}^n \mu_k f(t_k).$$

Since $f(t_k) = 0$,

$$\int_a^b \pi(t)p(t)\, dt = 0.$$

Conversely, assume that π has the same roots as y^{n+1}. Then π is a nonzero scalar multiple of y^{n+1}. Let f be an arbitrary polynomial of degree $\leq 2n - 1$. Dividing $f(t)$ by $\pi(t)$, we get $f(t) = q(t)\pi(t) + R(t)$, where the degrees of the quotient $q(t)$ and the remainder $R(t)$ are $\leq n - 1$. Thus, $f(t_k) = R(t_k)$, $(1 \leq k \leq n)$. Moreover, since

$$\int_a^b q(t)\pi(t)\, dt = 0,$$

$$\int_a^b f(t)\, dt = \int_a^b R(t)\, dt = \sum_{k=1}^n \mu_k f(t_k).$$

$$\text{Q.E.D.}$$

Corollary

The coefficients μ_k of the Gauss quadrature formula are positive.

Proof

Take i in $\{i, \cdots, n\}$. The polynomial

$$p(t) = \left(\frac{\pi(t)}{t - t_i}\right)^2$$

is of degree $2n - 2$. Thus,

$$\int_a^b p(t)\, dt = \sum_{k=1}^n \mu_k p(t_k).$$

Since $p(t_k) = 0$, if $k \neq i$, and $p(t_i) = (\pi'(t_i))^2$, then,

$$\int_a^b \left(\frac{\pi(t)}{t - t_i}\right)^2 dt = \mu_i(\pi'(t_i))^2;$$

whence $\mu_i > 0$.   Q.E.D.

We therefore have established the following theorem.

Theorem

For each n, let $\{t_k^n\}$ denote a Gauss set. Then

$$\lim_{n \to \infty} \left\{\sum_{k=1}^n \mu_k^n f(t_k^n)\right\} = \int_a^b f(t)\, dt \qquad \text{for every } f \in C[a, b].$$

B-3. The Conjugate of a Hilbert Space

An *isomorphism* between two normed linear spaces E and F is a one-to-one continuous linear operator $A: E \to F$ with $AE = F$. When such a map exists, the spaces E and F are called *equivalent*. An *isometry* A is a map from E into F such that $\|x\| = \|Ax\|$. If there exists a map from E to F, which is both an isomorphism and an isometry, then the spaces E and F are said to be *isometrically isomorphic*.

Our immediate aim is to show that Hilbert space H and its conjugate H^* are isometrically isomorphic.

Definitions

Two points x and y of H are *orthogonal* if $[x, y] = 0$. Similarly, two subspaces M and N of H are said to be orthogonal if $[M, N] = 0$. (This means that $[x, y] = 0$ for all $x \in M$ and $y \in N$.) The *orthocomplement* of a subset S of H is the set $\{x \in H : [x, S] = 0\}$. We denote it by $S^{\perp}$.

Lemma

Let S be a closed proper subspace of H. Then $S^{\perp}$ is closed and $H = S \oplus S^{\perp}$ (that is, every element z in H can be written uniquely as $x = y_0 + z_0$, with $y_0 \in S$ and $z_0 \in S^{\perp}$).

Proof

$S^{\perp} = \{x \in H: [x, y] = 0$, for all $y \in S\}$. For each $y \in S$, the set $\{x \in H: [y, x] = 0\}$ is closed because it is the inverse image of the closed set $\{0\}$ under the continuous function $[y, \cdot]$. Then, since $S^{\perp}$ is the intersection of these sets over $y \in S$, $S^{\perp}$ is closed. Since S is a proper subset, there is a point x in H but not in S; while, since S is closed and convex, there exists a closest point to x, say $y_0 \in S$. It will be shown that $x - y_0$ belongs to $S^{\perp}$. Let $y \neq 0$ be an arbitrary point of S. Then $y_0 + \alpha y$ is also in S for every number α; furthermore, $\|x - y_0 - \alpha y\| \geq \|x - y_0\|$. Thus, $-2\alpha[x - y_0, y] + \alpha^2 \|y\|^2 \geq 0$.

If $[x - y_0, y] < 0$, we replace y by $-y$, which also belongs to S. Thus, we may assume that $[x - y_0, y] \geq 0$. If the inner product is positive, the above inequality must be violated for small α. Thus, $x - y_0 = z_0$ belongs to $S^{\perp}$, and $x = y_0 + z_0$, with $y_0 \in S$ and $z_0 \in S^{\perp}$. If $x \in S$, $x = x + 0$; if $x \in S^{\perp}$, $x = 0 + x$, and if $x \in S \cap S^{\perp}$, then $[x, x] = 0$, which implies $x = 0$. Thus $H = S \oplus S^{\perp}$.

E X E R C I S E

Prove that the representation is unique.

Theorem (Frechet-Riesz)

For every $y^* \in H^*$, there exists a unique y in H such that $y^*(x) = [y, x]$ for all $x \in H$. The space H and H^* are isometrically isomorphic, and H^* is a Hilbert space.

Proof

If $y^* = 0$, set $y = 0$. If $y^* \neq 0$, let M denote the closed subspace $\{x \in H: y^*(x) = 0\}$, since $y^* \neq 0$, for some x, $y^*(x) \neq 0$. Hence, M is a proper subspace. By the preceding lemma, we may take a point $\hat{y} \neq 0$ in $M^\perp$. Set $y = \alpha \hat{y}$ with $\alpha = y^*(\hat{y})/\|\hat{y}\|^2$. Observe that $\alpha \neq 0$. Since $H = M \oplus M^\perp$, if $x \in H$, then $x = x_M^\perp + x_M$, with $x_M \in M$ and $x_{M_\perp} \in M^\perp$. Set

$$x_M = x - \beta \hat{y} \quad \text{and} \quad x_M^\perp = \beta y, \qquad \text{with } \beta = \frac{y^*(x)}{y^*(\hat{y})}.$$

Since $[y, x_M] = 0$, $[y, x] = \alpha[\hat{y}, \beta\hat{y}] = y^*(x)$, showing that y exists. If another point $\bar{y}$ had the same property, then $[y - \bar{y}, x] = 0$ for all $x \in H$. Take $x = y - \bar{y}$ and find that $y = \bar{y}$.

We have thus defined a mapping, say σ, from H^* into H. This mapping is 1-1, for given $y \in H$, the function $[y, \cdot]$ is a unique functional in H^*. If both y_1^* and y_2^* are mapped into y then $(y_1^* - y_2^*)(x) = 0$ for all x; hence, $y_1^* - y_2^*$ is the 0 functional. Since σ is linear on H^* ($\hat{y}$ is fixed), σ is an isomorphism. But by Schwarz's inequality,

$$\|y^*\| = \sup_{\|x\|=1} |y^*(x)| = \sup_{\|x\|=1} [y, x] = \|y\| = \|\sigma(y^*)\|.$$

Hence, σ is isometry as well. An inner product on H^* can be defined by $[y_1^*, y_2^*] = [\sigma(y_1^*), \sigma(y_2^*)]$, and by Section B-1, Remark 1, p. 106, H^* is complete.

B-4. The Frechet Differential

Let f be a map between normed-linear spaces E and F defined on an open subset E' of E. If, given $x \in E'$ and $\varepsilon > 0$, there exists $\delta > 0$ and a bounded linear operator $f'(x, \cdot)$ from E to F which is independent of ε and satisfies $\|f(x + h) - f(x) - f'(x, h)\| < \varepsilon\|h\|$, for all $x + h \in E'$ satisfying $\|h\| < \delta$, then f is said to be a *Frechet* or *F-differentiable at* x, and $f'(x, h)$ is called the *F-differential of* f *at* x. If $f'(x, h)$ exists for all x in a subset S of E', then f is said to be F-differentiable on S, and the map from S to $B(E, F)$, given by $f'(\cdot, \cdot)$, is called the F-derivative of f. The F-derivative and F-differential are also written as $f'(x)$ and $f'(x)h$, respectively. Here, $f'(x)$ or $f'(x, \cdot)$ is the value of the f derivative at x.

EXERCISES

1. Show that if f is a map from E_n to the reals with continuous partial derivatives that the Frechet differential $f'(x, h) = [\nabla f(x), h]$, where $\nabla f(x)$ is the gradient of f at x. Use the mean-value theorem and Schwarz's inequality.

2. Show that if f maps E_n into itself and the components f_i of f have continuous partials, then $f'(x, \cdot)$ is the linear transformation represented by the Jacobian matrix of f at x, with components $\partial f_i(x)/\partial x_j$, $(1 \leq i \leq n)$, $(1 \leq j \leq n)$.

3. Let x and y be functions defined on the reals with x and y' continuous. Define the map f from $C[0, 1]$ to the reals by

$$f(x) = \int_0^1 y(x(t)) \, dt.$$

Then

$$f'(x, h) = \int_0^1 y'(x(t))h(t) \, dt.$$

4. Show that when the F-derivative exists, it is unique.

5. If f is a bounded linear operator on E, show that $f'(x, \cdot) = f(\cdot)$ for all $x \in E$.

Let f be a real-valued function defined on a Hilbert space H. Assume that f has an F-differential $f'(x, h)$. The *gradient of f*, written $\nabla f(x)$, is a point of H satisfying $[\nabla f(x), h] = f'(x, h)$ for all h in H. The existence of $\nabla f(x)$ is guaranteed by Section B-3 (Frechet-Riesz theorem).

B-5. *The Gateaux Differential*

Let E and F be normed linear spaces. Given x and h in E and $\varepsilon > 0$, let f be a map defined on $\{x + \lambda h : |\lambda| < \varepsilon\}$, taking values in F. The map f is said to have a G-differential (G = Gateaux) at x with increment h if the limit

$$\lim_{\lambda \to 0} \left(\frac{1}{\lambda}\right)(f(x + \lambda h) - f(x)) = f'(x, h)$$

exists in F. If f has a G-differentiable at x for every h in E, then f is said to be G-differentiable at x.

Similarly, if $f'(\cdot, h)$ is a map from $\{x + \lambda k : |\lambda| < \varepsilon\}$ to F for fixed h, and

$$\lim_{\lambda \to 0} \left(\frac{1}{\lambda}\right)(f'(x + \lambda k, h) - f'(x, h)) = f''(x, h, k)$$

exists, f has a second G-differential at x with increments h and k, and is twice G-differentiable if this limit exists for every h and k.

Let S be a subset of E. The G-differential $f'(x, h)$ is said to be bounded on S if there exists a constant β such that $\|f'(x, h)\| \leq \beta\|h\|$, for all h in E and x in S. Similarly, if f is twice G-differentiable, $f''(x, h, k)$ is bounded on S if for all h and k in E and x in S, there exists β' such that $\|f''(x, h, k)\| \leq \beta'\|h\|\|k\|$.

EXERCISES

1. Show that $f'(x, \mu h) = \mu f'(x, h)$ and $f''(x, \mu h, \lambda k) = \mu\lambda f''(x, h, k)$.
2. Show that F-differentiability implies G-differentiability.

B-6.   The "Chain Rule"

Given normed linear spaces E, F, and G, let $f: E \to F' \subset F$, $g: F' \to G$, and let h denote the composition gf. Assume that f is F-differentiable at z in E, that the range of f is an open subset F' of F, and that g is F-differentiable at $f(z) = x$ in F. Then $h'(z, \cdot) = g'(x, f'(z, \cdot))$.

Proof

Set $M = \|f'(z, \cdot)\| + 1$. Given $\varepsilon > 0$, choose

$$\varepsilon_1 = \min\left\{ 1, \frac{\varepsilon}{(M + \|g'(x, \cdot)\|)} \right\}.$$

Choose δ_0 such that $\|t\| < \delta_0$ implies that

$$\|f(z + t) - f(z) - f'(z, t)\| < \varepsilon_1\|t\|.$$

Choose δ_1 such that if $y = f(z + t)$ and $\|y - x\| < \delta_1$, then

$$\|g(y) - g(x) - g'(x, y - x)\| < \varepsilon_1\|y - x\| \tag{a}$$

Choose $\delta = \min\{\delta_0, \delta_1/M\}$. Choose $\|t\| < \delta$. Then $\|y - x\| < \delta_1$ because $\|f(z + t) - f(z)\| \leq \varepsilon_1\|t\| + \|f'(z, \cdot)\|\|t\| \leq M\|t\| \leq \delta_1$.

By (a), $\|h(z + t) - h(z) - g'(x, y - x)\| < \varepsilon_1\|y - x\| < \varepsilon_1 M\|t\|$. Calculate that

$$\|g'(x, y - x) - g'(x, f'(z, t))\|$$
$$= \|g'(x, f(z + t) - f(z) - f'(z, t)\| \le \|g'(x, \cdot)\|\varepsilon_1\|t\|.$$

Adding this to (a), we get

$$\|h(z + t) - h(z) - g'(x, f'(z, t))\| \le \varepsilon_1\|t\|[M + \|g'(x, \cdot)\|] \le \varepsilon\|t\|.$$

Since this holds whenever $\|t\| < \delta$, $g'(x, f'(z, \cdot)) = h'(z, \cdot)$. Q.E.D.

B-7. Taylor's Formula for Twice G-Differentiable Real-Valued Functions

Let $[a, b]$ be an interval of reals and g a real-valued function such that g'' is defined on $[a, b]$. Then, since g' is continuous on $[a, b]$ the Taylor series

$$g(\gamma) = g(\gamma_0) + (\gamma - \gamma_0)g'(\gamma_0) + g''(\theta)(\gamma - \gamma_0)^2$$

is valid when γ and γ_0 belong to $[a, b]$ and θ is some number satisfying $\theta = \lambda\gamma + (1 - \lambda)\gamma_0$, $\lambda \in (0, 1)$.[4]

Theorem

Let f be a real-valued function defined on a linear space E. Assume that x and h belong to E and that f has two G-differentials with increment h at each point of the segment $\{x + \gamma h: \gamma \in [0, 1]\}$. Then, if γ and γ_0 are in $[0, 1]$,

$$f(x + \gamma h) = f(x + \gamma_0 h) + f'(x + \gamma_0 h, (\gamma - \gamma_0)h)$$
$$+ \tfrac{1}{2}f''(x + \theta h, (\gamma - \gamma_0)h, (\gamma - \gamma_0)h)$$

with $\theta = t\gamma + (1 - t)\gamma_0$ for some $t \in (0, 1)$.

Proof

Let $g(\gamma) = f(x + \gamma h)$. If $\gamma_0 \in [0, 1]$, we observe that

$$g'(\gamma_0) = f'(x + \gamma_0 h, h), \quad g''(\gamma_0) = f''(x + \gamma_0 h, h, h),$$

and $g''(\gamma_0)$ is defined for all $\gamma_0 \in [0, 1]$. Thus, if $\gamma \in [0, 1]$,

$$g(\gamma) = g(\gamma_0) + (\gamma - \gamma_0)g'(\gamma_0) + \frac{(\gamma - \gamma_0)^2}{2}g''(\theta)$$

$$= f(x + \gamma h) = f(x + \gamma_0 h + (\gamma - \gamma_0)h)$$

$$= f(x + \gamma_0 h) + (\gamma - \gamma_0)f'(x + \gamma_0 h, h) + \frac{(\gamma - \gamma_0)^2}{2}f''(x + \theta h, h, h),$$

---

[4] See, e.g., T. M. Apostol, *Mathematical Analysis*, Addison-Wesley Publishing Company, Reading, Mass., 1957, p. 96.

with $\theta = t\gamma + (1 - t)\gamma_0$ for some $t \in (0, 1)$. The theorem follows by making use of the Problem of Section B-5. If we set $\gamma = 1$ and $\gamma_0 = 0$, we get the usual form of the Taylor series: $f(x + h) = f(x) + f'(x, h) + \frac{1}{2}f''(x + \theta h, h, h)$.

PROBLEM

Assume that f is real-valued and G-differentiable at x. If $f'(\cdot, h)$ is continuous for each fixed h in an open set containing x, then $f'(x, \cdot)$ is linear. That is, $f'(x, \alpha h + \beta k) = \alpha f'(x, h) + \beta f'(x, k)$. *Hint:* Use the mean-value theorem: $f(x + h) - f(x) = f'(x + \theta h, h)$.

Prove the same theorem when f takes its values in a Banach space. *Hint:* Assume that the following makes sense:

$$\int_0^1 \lim_{\mu \to 0} \frac{f(x + (t + \mu)h) - f(x + th)}{\mu}\, dt = \int_0^1 f'(x + th, h)\, dt$$

$$= \int_0^1 \frac{d}{dt} f(x + th)\, dt = f(x + h) - f(x).$$

B-8. Weak Convergence

A sequence $\{x_k\}$ in a Banach space E is said to *converge weakly* to z, written: $x_k \rightharpoonup z$ iff $f(x_k) \to f(x)$ for all $f \in E^*$. Clearly, convergence or *strong convergence*, as it is called to distinguish it from weak convergence, *implies weak convergence*.

Example

Consider the Hilbert space l_2 of square, summable sequences. The sequence $\{x^1, x^2, \cdots\} = \{(1, 0, 0, \cdots), (0, 1, 0, 0, \cdots) \cdots\}$ converges weakly to 0 because if we identify functionals f in l_2^* with points in l_2 of the form $(f_1, f_2, \cdots)$, we must have that $\lim_{n \to \infty} f_n = 0$. (Otherwise, $\sum_{i=1}^{\infty} f_i^2$ would diverge.) Thus, for each $f \in l_2^*$, $f(x^n) = [f, x^n] = f_n$. Thus, $\{f(x^n)\} \to 0 = f(0)$ for all $f \in l_2^*$, and consequently $\{x^n\} \rightharpoonup 0$. Observe, however, that $\|x^n\| = 1$ for all n. This shows us that the weak convergence does not imply strong convergence. On the other hand, we have the result of the following exercise.

EXERCISE

Prove that if E is finite dimensional, weak and strong convergence are equivalent.

Hint: Choose a basis, say $(e_1, \cdots, e_n)$, so that

$$x \in E \Rightarrow x = x_1 e_1 + x_2 e_2 + , \cdots, x_n e_n.$$

Consider the functionals $f_i \in E^*$, $(1 \le i \le n)$, for which $f_i(e_i) = 1$ and $f_i(e_k) = 0$, $k \ne i$.

In the sequel we introduce the concepts of weak closure, continuity, and compactness. More precisely, the concepts are weak *sequential* closure,* continuity, and compactness if we wish to adhere to the standard vernacular. For the sake of brevity (and euphony), however, we shall suppress the word "sequential."

A function Q mapping a B-space E into a B-space F will be called *weakly continuous* if $\{x_k\} \rightharpoonup x$ in E implies $\{Qx_k\} \rightharpoonup Qx$. If F is the space of reals or finite dimensional space, equivalently $\{Qx_k\} \to Qx$.

Example

Let A be a bounded linear operator from E to F; then A is weakly continuous. To see this, let f denote an arbitrary functional in F^* and let $\{x_n\} \rightharpoonup x$. Then $f(Ax_n) = g(x_n)$, where $g \in E^*$. Hence, $\{g(x_n)\} \to g(x)$ and $\{f(Ax_n)\} \to f(Ax)$. This shows that $\{Ax_n\} \rightharpoonup Ax$, since f was arbitrary in F^*.

A set S in a B-space E is *weakly closed* if $\{x_k\} \rightharpoonup x$ and $x_k \in S$ implies that x is in S. If C is weakly closed, it is strongly closed. The converse is not always true, but we do have the following theorem.

Theorem

A closed convex subset C of Hilbert space H is weakly closed.

Proof

Let $\{x_k\}$ be a sequence in C converging weakly to x. If $x \notin C$, we have by the separation theorem $[x_k - P(x), x - P(x)] \le 0$, while

$$[x - P(x), x - P(x)] > 0.$$

Identity $f \in H^*$ with $x - P(x)$. Then $f(x_k) \le f(P(x)) < f(x)$, contradicting that x_k converges weakly to x.

A weakly closed subset S of H is weakly compact if every sequence in S has a weakly converging subsequence. A real-valued function defined on H is *weakly* l.s.c. if $\{x_k\} \rightharpoonup x$ implies that $\lim \inf \{f(x_k)\} \ge f(x)$.

Lemma

A continuous convex function f defined on a closed convex subset C of H is weakly l.s.c.

---

* Weak sequentially closed sets are not necessarily closed in the "weak topology." We shall however only need weak sequential closure, weak sequential continuity, etc.

Proof

The set $S = \{x \in C : f(x) \le \mu\}$ is closed because f is continuous; it is convex because f is convex. Thus, as above, S is weakly closed for all M. By the same argument as Section D-5, Chapter I, after replacing the strongly convergent sequence by weakly convergent one, we find that f is weakly l.s.c.

Remark

Arguing as in Chapter I, Section D-5, p. 41, we see that a weakly l.s.c. function defined on a weakly compact set achieves its minimum thereon.

B-9. Weak Compactness Theorem

Bounded and weakly closed subsets of Hilbert space H are weakly compact.

Proof

The proof is deferred to Section C-3, p. 140.

E X E R C I S E

Prove that the distance from a point to a closed convex subset of Hilbert space is achieved without using the parallelogram equality. Use instead the notions of weak l.s.c. and weak compactness.

Lemma

Let E be a convex subset of a normed linear space and f a real-valued function defined on E. Assume that f is twice G-differentiable, and that for every x and h in E there exists a number $\mu \ge 0$ such that $f''(x, h, h) \ge \mu\|h\|^2$. Then f is convex on E.

Proof

Let x and y be given in E. We show that $g(\theta) = (1 - \theta)f(x) + \theta f(y) - f((1 - \theta)x + \theta y)$ is nonnegative if $0 \le \theta \le 1$. Let $u = (1 - \theta)x + \theta y$; then

$$g(\theta) = \theta(f(y) - f(x)) + f(x) - f(u)$$

$$= (1 - \theta)(f(x) - f(u)) + \theta(f(y) - f(u))$$

$$= (1 - \theta)[f'(u, x - u) + \tfrac{1}{2}f''(\xi, x - u, x - u)]$$

$$+ \theta[f'(u, y - u) + \tfrac{1}{2}f''(\eta, y - u, y - u)],$$

where ξ and η belong to the open segment joining x and y, if $0 < \theta < 1$. Because $x - u = \theta(x - y)$ and $y - u = -(1 - \theta)(x - y)$,

$$g(\theta) = \tfrac{1}{2}(1 - \theta)\theta^2 f''(\xi, x - y, x - y) + \tfrac{1}{2}\theta(1 - \theta)^2 f''(\eta, x - y, x - y)$$
$$\geq \tfrac{1}{2}\theta(1 - \theta)\mu\|x - y\|^2 \geq 0.$$

Remark

It is clear that if $\mu > 0$, then f is strictly convex.

B-10. *Characterization of Extremals in Convex Programming*

Lemma

Let C be a closed convex subset of Hilbert space H, with $p \in H$, $r \in C$ closest to p, and $q \in C$. Then

$$[p - q, r - q] \geq \|r - q\|^2.$$

Proof

If $p \in C$, the inequality is trivially true. If $p \notin C$, the set C is supported by a hyperplane through r with normal $p - r$; thus,

$$[p - q, r - q] = [r - q + p - r, r - q]$$
$$= \|r - q\|^2 + [p - r, r - q] \geq \|r - q\|^2.$$

Theorem

Let f be an F-differentiable convex function defined on a closed convex set C in a Hilbert space. The following statements are equivalent:

(i) $z = P(z - \nabla f(z))$;
(ii) z minimizes $[\nabla f(z), \cdot]$ on C.
(iii) z minimizes f on C.

Here P denotes the projection map for C (p. 100).

Proof

To prove (i) $\Rightarrow$ (ii), we observe that if $\nabla f(z) = 0$, the conclusion is trivial. Assume, then, that $\nabla f(z) \neq 0$ and observe that $z - \nabla f(z) \notin C$ because $P(y) = y$ for all $y \in C$. By hypothesis, z is closest in C to $z - \nabla f(z)$; so by the separation theorem, $[-\nabla f(z), x - z] \leq 0$ for all $x \in C$, verifying statement (ii). To prove (ii) $\Rightarrow$ (i), observe that by (ii), C lies on the "negative" side of the hyperplane

$\{x: [-\nabla f(z), x - z] = 0\}$. The closest point in H to $z - \nabla f(z)$ is the projection of $z - \nabla f(z)$ on H, which is z. No other point in C can be closer than z to $z - \nabla f(z)$ because if it were, it must lie on the "positive" side of H.

We prove next that (ii) $\Rightarrow$ (iii). Assume that (iii) is false and choose $x \in C$ such that $f(x) < f(z)$. Then $f(z + \lambda(x - z)) \le f(z) + \lambda(f(x) - f(z))$ by the convexity of f. Thus, $\lambda^{-1}(f(z + \lambda(x - z)) - f(z)) \le f(x) - f(z) < 0$. By Section B-5, the G-differential $f'(z, x - z)$ exists. Evidently it is negative. On the other hand, by Section B-5, the G-differential $f'(z, x - z)$ is equal to the F-differential $[\nabla f(z), x - z]$ which by (ii) is nonnegative. This contradiction establishes (iii).

We now show that (iii) $\Rightarrow$ (ii). Suppose that (ii) is false; then $[\nabla f(z), x] < [\nabla f(z), z]$ for some $x \in C$. Since C is convex, the open segment $\langle x, z \rangle$ belongs to C. Since $x \to [\nabla f(z), x]$ is linear, every point u in $\langle x, z \rangle$ satisfies

$$[\nabla f(z), u] < [\nabla f(z), z].$$

Since f is differentiable at z, $|f(u) - f(z) - f'(z, u - z)| < \varepsilon \|u - z\|$, provided $\|u - z\| < \delta$. Choose ε so that $\varepsilon < -[\nabla f(z), (x - z)/\|x - z\|]$ and u so that $\|u - z\| < \delta$. Then

$$f(u) < f(z) + [\nabla f(z), u - z] + \varepsilon \|u - z\|$$

$$= f(z) + \|u - z\| \left[\varepsilon + \left[\nabla f(z), \frac{u - z}{\|u - z\|} \right] \right] < f(z),$$

a contradiction.

Remark

(iii) $\Rightarrow$ (i) does not use the convexity of f.

Lemma

Assume f not necessarily convex. If there exists a positive number β and a neighborhood $N(z)$ such that the second G-differential of f exists and satisfies $f''(\xi, h, h) \le \beta \|h\|^2$, if $\xi \in N(z)$, or if x and y in $N(z)$ imply that $\|\nabla f(x) - \nabla f(y)\| \le \beta \|x - y\|/2$, then in the notation below, $f(z) - f(z(\gamma)) \ge \|h(\gamma)\|^2 \theta(\gamma)$.

Proof

To prove that (iii) $\Rightarrow$ (i), observe first that if $P(z - \gamma \nabla f(z)) = z$ for some $\gamma > 0$, then $P(z - \gamma \nabla f(z)) = z$ for all $\gamma > 0$. Assume (i) to be false. Thus, $P(z - \nabla f(z)) \ne z$, and $P(z - \gamma \nabla f(z)) \ne z$ for all $\gamma > 0$. Set $P(z - \gamma \nabla f(z)) = z(\gamma)$ and use the above lemma with $p = z - \gamma \nabla f(z)$, $q = z$, and $r = z(\gamma)$. Thus,

$$[-\gamma \nabla f(z), z(\gamma) - z] \ge \|z(\gamma) - z\|^2.$$

Let $h(\gamma) = z(\gamma) - z$. By Taylor's theorem, $f(z) - f(z(\gamma)) = -[\nabla f(z), h(\gamma)] - \frac{1}{2}f''(\xi(\gamma), h(\gamma), h(\gamma))$, where $\xi(\gamma)$ lies "between" z and $z(\gamma)$. Thus,

$$f(z) - f(z(\gamma)) \geq \|h(\gamma)\|^2 \left[\frac{1}{\gamma} - \frac{\beta}{2}\right] = \|h(\gamma)\|^2 \theta(\gamma)$$

whenever γ is sufficiently small so that $\xi(\gamma) \in N(z)$. The continuity of P guarantees that $\lim_{\gamma \to 0} z(\gamma) = z$, which in turn guarantees that for some γ, $\xi(\gamma) \in N(z)$. Furthermore, γ may be chosen so small that $1/\gamma > \beta$, showing that $\theta(\gamma) > 0$.

B-11. Convex Programming

Our object in this section is to generalize the results of Chapter I, Section D-2, in two directions: first to minimization, constrained to a convex set rather than to the entire space; second to infinite dimensional Hilbert space, rather than to finite dimensional.

In what follows, C will denote a closed convex subset of Hilbert space H, and P will be the projection map for C. We denote by f a real-valued function on H, and by S, the set $\{x \in C : f(x) \leq f(x^0)\}$, where the point x^0 is arbitrary in C. Denote by $\hat{S}$ any open set containing the convex hull of S. As in Section B-4, Exercise 1, we denote by $\nabla f(x)$ the gradient of f; that is to say, the representor in H of the bounded linear functional $f'(x, \cdot)$, which is the F-derivative of f at x. A point z in C will be called a *stationary point* if $P(z - \nabla f(z)) = z$.

Theorem

Assume that f is bounded below. For each $x \in \hat{S}$, $h \in H$, and for some $\gamma_0 > 0$, assume that the F-differential $f'(x, h)$ exists and that the G-differential $f''(x, h, h)$ exists and satisfies

$$f''(x, h, h) \leq \frac{\|h\|^2}{\gamma_0}.$$

Choose δ and γ_k satisfying $0 < \delta \leq \gamma_0$ and $\delta \leq \gamma_k \leq 2\gamma_0 - \delta$. Let $x_{k+1} = P(x_k - \gamma_k \nabla f(x_k))$. Then:

(i)   The sequence $\{x_k\}$ belongs to S, $\{(x_{k+1} - x_k)\}$ converges to 0, and $\{f(x_k)\}$ converges downward to a limit L.

(ii) If S is compact, z is a cluster point of $\{x_k\}$, and ∇f is continuous in some neighborhood of z, then z is a stationary point. If z is unique, $\{x_k\}$ converges to z, and z minimizes f on C.

(iii)   If S is convex, ∇f is bounded on S, and $f''(x, h, h) \geq \mu\|h\|^2$ for each $x \in S$, $h \in H$, and some $\mu \geq 0$, then $L = \inf\{f(x): x \in C\}$.

(iv)   Assume (iii) with S bounded. Weak cluster points of $\{x_k\}$ minimize f on C.

(v)   Assume (iii) with μ positive and ∇f bounded on S. Then $f(z) = L$ for some z in S, $\{x_k\}$ converges to z, and z is unique.

Proof

Assume that $x_k \in S$ and that x_k is not stationary. Set

$$\nabla f(x_k) = \nabla f_k, \qquad x(\gamma) = P(x_k - \gamma \nabla f_k), \qquad u_k(\gamma) = x(\gamma) - x_k,$$

and

$$\Delta(x_k, \gamma) = f(x_k) - f(x(\gamma)).$$

Recalling the proof of the corollary in Section B-10, we may write

$$-\gamma[\nabla f_k, u_k(\gamma)] \geq \|u_k(\gamma)\|^2;$$

and, provided γ is sufficiently small,

$$\Delta(x_k, \gamma) \geq \|u_k(\gamma)\|^2 \left\{ \gamma^{-1} - \frac{f''\,(\xi(\gamma), u_k(\gamma), u_k(\gamma))}{2\|u_k(\gamma)\|^2} \right\}.$$

Here $\xi(\gamma) = x_k + tu_k(\gamma)$ with $t \in (0, 1)$. Let $\hat{\gamma}$ denote the least positive γ satisfying $\Delta(x_k, \gamma) = 0$, if such exists. If $\hat{\gamma}$ does not exist, $f(x(\gamma)) < f(x_k)$ for all $\gamma > 0$. If $\hat{\gamma}$ exists, then

$$\Delta(x_k, \hat{\gamma}) = 0 \geq \frac{2\|u_k(\hat{\gamma})\|^2}{f''(\xi(\hat{\gamma}), u_k(\hat{\gamma}), u_k(\hat{\gamma}))} - \hat{\gamma},$$

whence $\hat{\gamma} \geq 2\gamma_0$. Thus, if $\delta \leq \gamma \leq 2\gamma_0 - \delta$, $\Delta(x_k, \gamma) > 0$ and $x(\gamma) \in S$, and therefore

$$\Delta(x_k, \gamma_k) \geq \frac{\|x_{k+1} = x_k\|^2 \delta}{4\gamma_0^2},$$

proving (i).

To prove (ii), we write

$$\|x_{k+1} - x_k\| = \|P(x_k - \gamma_k \nabla f_k) - x_k\|,$$

and take a subsequence $\{x_k\}$ such that $\{x_k\} \to z$ and $\gamma_k \to \gamma^*$. Then, by continuity,

$$\{P(x_k - \gamma_k \nabla f_k) - x_k\} \to (P(z - \gamma^* \nabla f(z)) - z) = 0,$$

showing that z is a stationary point. Observe that if $\{x_k - \gamma_k \nabla f_k\} \in C$ infinitely often, then $\{\nabla f(x_k)\} \to 0$. Since f achieves a minimum on C, it does so at a stationary point. (See Corollary, Section B-10.) If there is only one stationary point, then it minimizes f, and $\{x_k\}$ converges to it, which proves (ii).

To prove (iii), assume that $L \neq \inf\{f(x): x \in C\}$ and choose $z \in C$ such that $f(z) < L$. By Taylor's theorem, $f(z) \geq f(x_k) + [\nabla f_k, z - x_k]$. If

$$\lim \inf [\nabla f_k, z - x_k] = B$$

were nonnegative, then $f(z) \geq \lim \inf [f(x_k) + [\nabla f_k, x_k]] \geq L$, a contradiction. Observe that either $x_k - \gamma_k \nabla f_k - x_{k+1}$ is the normal to C at x_{k+1} or it is 0. Thus, in either event,

$$[x_k - \gamma_k \nabla f_k - x_{k+1}, z - x_{k+1}] \leq 0;$$

or equivalently,

$$[\gamma_k \nabla f_k, z - x_k] \geq [\gamma_k \nabla f_k, x_{k+1} - x_k] + [x_k - x_{k+1}, z] + [x_{k+1} - x_k, x_{k+1}]$$

$$\geq -2\gamma_0 \|\nabla f_k\| \|x_{k+1} - x_k\| + [x_k - x_{k+1}, z]$$

$$+ [x_{k+1} - x_k, x_{k+1}].$$

If $\{x_k\}$ is bounded, clearly $B = 0$. Otherwise, take a subsequence satisfying $\|x_{k+1}\| > \|x_k\|$. Then $\|x_{k+1}\|^2 > \|x_{k+1}\| \|x_k\|$ and $[x_{k+1} - x_k, x_{k+1}] > 0$, whence $B \geq 0$.

To prove (iv), we recall by Section B-8 (lemma) that a continuous convex function defined on a closed convex subset of H is weakly l.s.c. Thus, if $\{x_k\}$ converges weakly to z, then $\lim \inf (f(x_k)) = L \geq f(z) \geq L$.

To prove (v), we assume that $s > k$ and use Taylor's theorem to write

$$0 > f(x_s) - f(x_k) \geq [\nabla f_k, x_s - x_k] + \frac{\mu \|x_s - x_k\|^2}{2}$$

or

$$0 > -\|\nabla f_k\| \|x_s - x_k\| + \frac{\mu}{2} \|x_s - x_k\|^2.$$

If $\{x_s\}$ is unbounded and k is fixed, we arrive at a contradiction because the right-hand side of this inequality tends to ∞. Thus, $\{x_s\}$ is bounded. Again using the fact that $x_k - \gamma_k \nabla f_k - x_{k+1}$ is normal to x_{k+1} or is 0, we write

$$[\gamma_k \nabla f_k, x_s - x_k] \geq [\gamma_k \nabla f_k, x_{k+1} - x_k] + [x_k - x_{k+1}, x_s - x_{k+1}].$$

Thus, $0 > [\gamma_k \nabla f_k, x_s - x_k] \geq 0$, while $\lim \inf [\gamma_k \nabla f_k, x_s - x_k] \geq 0$. Since

$$\|x_s - x_k\|^2 < \frac{2[f(x_k) - f(x_s) - [\nabla f_k, x_s - x_k]]}{\mu},$$

it follows that $\{x_s\}$ is a Cauchy sequence. There exists, therefore, z in S minimizing f on C. By the theorem of Section B-10, $[\nabla f(z), x - z] \geq 0$. Thus, by Taylor's theorem,

$$f(x) \geq f(z) + \frac{\mu \|x - z\|^2}{2},$$

and therefore z is unique.

Remarks

(a) In Section C-4, Remark 1, it will be seen that

$$f''(x, h, h) \leq \frac{1}{\gamma_0} \|h\|^2$$

implies that ∇f is Lipschitz continuous on S. Thus, the assumption in (ii) that ∇f is continuous in a neighborhood of z is unnecessary. In Section C-5, Exercise, p. 153, one can show that if ∇f is uniformly continuous on a bounded set, say S, then ∇f is bounded on S and f is bounded below on S.

(b) Observe that the completeness of H enters three times: first, to guarantee the existence of the projection on C; secondly in (iv); and for the third time in (v) to guarantee that the Cauchy sequence has its limit.

B-12. Rate of Convergence

In this section we shall discuss the rate of convergence of the sequence $\{x_k\}$ defined in the theorem of Section B-11 in a very special case. We shall assume the hypothesis of (v), that $C = H$ and that the second F-differential exists at the minimum point z. In order to carry out this proof, it is necessary to examine several new concepts. We turn our attention again to Frechet differentiation.

The Second F-Differential

Let E and F be normed linear spaces and $f: E \to F$. If the F-differential $f'(x, h)$ exists, recall that $f'(x, h) \in F$, $f'(x, \cdot) \in B(E, F)$ (the normed linear space of bounded linear operators from E to F), and $f'(\cdot, \cdot) = f': E \to B(E, F)$. Since $B(E, F)$ is a normed linear space, say $\hat{F}$, we have at hand a new map, $f': E \to \hat{F}$. If the map f' is F-differentiable, we call its F-derivative f''. Since f'' is the F-derivative of the map f', we have that $f'': E \to B(E, \hat{F}) = B(E, B(E, F))$. Thus, the value of f'' at x is a bounded linear operator from E to $B(E, F))$. Applying the definition of the F-differential to the mapping

$x \to f'(x, \cdot)$, we see that f is twice F-differentiable at x in E if, given $\varepsilon > 0$, there exists $\delta > 0$ and a bounded linear operator $f''(x, \cdot, \cdot)$ independent of ε in $B(E, B(E, F))$ such that $\|f'(x + k, \cdot) - f'(x, \cdot) - f''(x, \cdot, k)\| < \varepsilon \|k\|$ for all k such that $\|k\| < \delta$. Since $f''(x, \cdot, k)$ is the F-differential of $f'(x, \cdot)$, this differential is unique whenever it exists, and therefore so is $f''(x, \cdot, \cdot)$. Take (h, k) arbitrarily in $E \times E$. The point $f''(x, h, k)$ in F is called the *second F-differential of f*. Observe that the following conditions are direct consequences of the definition of $f''(x, \cdot, \cdot)$:

(i) $f''(x, \cdot, k)$ is linear and bounded in the variable k.
(ii) For fixed $h \in E$, $f''(x, h, k)$ is linear and bounded in the variable k.
(iii) Because $f''(x, \cdot, k) \in B(E, F)$, $f''(x, h, k)$ is linear and bounded in h for fixed k.
(iv) There exists a constant $\mu > 0$ such that $\|f''(x, h, k)\| \le \mu \|h\| \|k\|$, where μ is the norm of the operator $f''(x, \cdot, \cdot)$.

To prove (iv), we have that since $f''(x, \cdot, \cdot) \in B(E, B(E, F))$, it follows for all $k \in E$ that $\|f''(x, \cdot, k)\| \le \mu \|k\|$ for some $\mu > 0$. Here the left norm is taken in the space $B(E, F)$ and the right norm in the space E. Thus,

$$\|f''(x, \cdot, k)\| = \sup\left\{ \left\| f''\left(x, \frac{h}{\|h\|}, k\right) \right\| : h \in E \right\},$$

the norm enclosing the second differential being the norm in F. Thus, for all (h, k), in $E \times E$, $\|f''(x, h, k)\| \le \mu \|h\| \|k\|$.

Self-adjoint Operators in H

Let A be a bounded linear operator mapping a Hilbert space H into itself. *The adjoint A^* of A is a bounded linear operator that satisfies* $[Ax, y] = [x, A^*y]$ for all (x, y) in $H \times H$. As an example, if A is an $n \times n$ matrix mapping the Hilbert space E_n into itself and A^T is the transpose of A, then $A^T = A^*$ because $[Ax, y] = x^T A^T y = [x, A^T y]$. The operator A is said to be *self-adjoint if $A^* = A$*. The numbers $\mu = \inf\{[x, Ax] : \|x\| = 1\}$ and $\lambda = \sup\{[x, Ax] : \|x\| = 1\}$ are called *bounds for the spectrum* of the self-adjoint operator A.

Lemma

Assume that A is a self-adjoint operator on H.

(a) $\sup\{|[x, Ax]| : \|x\| = 1\} = \sup\{\|Ax\| : \|x\| = 1\} = \|A\|$.
(b) If I is the identity and μ and λ are the spectral bounds for A, then $\|I - \gamma A\| \le \max\{|1 - \gamma\mu|, |1 - \gamma\lambda|\}$.

Proof

See Chapter I, Section C-4, p. 24

Lemma

Let f be a map from a B-space to the reals. Assume that f has a second F-differential at x in E. Then $f''(x, h, k) = f''(x, k, h)$ for all (h, k) in $E \times E$.

Proof

Observe that

$$f(x + th) - f(x) = \int_0^1 \frac{d}{ds} f(x + sth) \, ds$$

$$= \int_0^1 f'(x + sth, th) \, ds.$$

Let

$$g(t, h, k) = t^{-2}(f(x + th + tk) - f(x + tk)) - t^{-2}(f(x + th) - f(x))$$

$$= \int_0^1 t^{-1}[f'(x + sth + tk, h) - f'(x + sth, h)] \, ds.$$

Thus,

$$g(t, h, k) - f''(x, h, k)$$

$$= \int_0^1 \{t^{-1}[f'(x + sth + tk, h) - f'(x, h)] - f''(x, h, sh + k)\} \, ds$$

$$- \int_0^1 \{t^{-1}[f'(x + sth, h) - f'(x, h)] - f''(x, h, sh)\} \, ds.$$

Since f' is F-differentiable at x, given $\varepsilon > 0$, there exists $\delta > 0$ such that $\|t(sh + k)\| < \delta$ implies

$$\left| \frac{f'(x + t(sh + k), h) - f'(x, h)}{t} - f''(x, h, (sh + k)) \right| < \varepsilon \|sh + k\| \, \|h\|.$$

By choosing t sufficiently small, this inequality can be satisfied with ε arbitrarily small. Similarly, the remaining integrand can be bounded by $\varepsilon \|sh\| \, \|h\|$. It follows that $\lim_{t \to 0} g(t, h, k) = f''(x, h, k)$. The proof follows because $g(t, \cdot, \cdot)$ is symmetric for all $t \neq 0$.

Corollary

Assume that f is twice differentiable at x in H. There exists a bounded linear self-adjoint operator $Q(x)$ from H to H such that $f''(x, h, k) = [Q(x)k, h]$ for all (h, k) in $H \times H$. We call $Q(x)$ *the Hessian of f* at x.

Proof

Recall that if $f \colon E \to F$, the value of f'' at x is a bounded linear operator from E to $B(E, F)$. In the case at hand, the map is from H to H^*. Let σ denote the linear mapping from H^* to H that is constructed in the Frechet-Riesz theorem, and verify that $Q(x) = \sigma f''(x, \cdot, \cdot)$ is a bounded linear map from H to itself. Since $f''(x, h, k) = [Q(x)k, h]$ for all (h, k) in $H \times H$ and $f''(x, \cdot, \cdot)$ is symmetric, $Q(x)$ is self-adjoint.

Theorem

Assume the hypothesis of (v) of the theorem of Section B-11. Assume that $C = H$ and that the second F-differential exists at the minimizer z. Then the sequence $\{x_k\}$ constructed in Section B-11 converges to z at a speed greater than some geometric progression.

Proof

Let λ denote the upper bound for the spectrum of the Hessian $Q(z)$. For $\gamma \in J = [\delta, 2\gamma_0 - \delta]$, define

$$\beta(\gamma) = \max\{|1 - \gamma\mu|, |1 - \gamma\lambda|\} \geq \|I - \gamma Q(z)\|.$$

Set $\bar{\beta} = \max\{\beta(\gamma) : \gamma \in J\}$. Then $\bar{\beta} < 1$. Choose $\varepsilon = (1 - \bar{\beta})/4\gamma_0$. Thus,

$$\|x_{k+1} - z\| = \|x_k - z - \gamma_k \Delta f_k\|$$

$$\leq \|x_k - z - \gamma_k Q(z)(x_k - z)\| + \gamma_k \|\nabla f_k - Q(z)(x_k - z)\|.$$

Since f'' exists at z, we have that whenever $\|x_k - z\| < \delta$,

$$\|f'(x_k, \cdot) - f'(z, \cdot, x_k - z) - Q(z)(x_k - z)\| < \varepsilon\|x_k - z\|.$$

Because $\{x_k\}$ converges to z, we can choose K sufficiently large so that $k \geq K$ implies $\|x_k - z\| < \delta$. Since $f'(z, \cdot) = 0$,

$$\|\nabla f(x_k) - Q(z)(x_k - z)\| < \varepsilon\|x_k - z\|.$$

Thus,

$$\|x_{k+1} - z\| \leq (\|I - \gamma_k Q(z)\| + \varepsilon\gamma_k)\|x_k - z\| \leq \frac{\|x_k - z\|(1 + \bar{\beta})}{2}.$$

Q.E.D.

Lemma

Let A be a continuous nonlinear operator defined throughout H with values in H. Given x_0 arbitrarily in H, set

$$f(x) = \int_0^1 [A(x_0 + t(x - x_0)), x - x_0] \, dt \quad \text{and} \quad S = \{x \in H : f(x) < 0\}.$$

Assume that A is F-differentiable on $H(S)$ and set $A'(x, h) = Q(x)h$. For $x \in S$, assume that Q is continuous and $Q(x)$ is self-adjoint. Then $A(x) = \nabla f(x)$.

Proof

We show that $x \in S$ and $h \in H$ imply that $f'(x, h) = [A(x), h]$. Observe that

$$f'(x, h) = \int_0^1 \{[A'(x_0 + t(x - x_0)), th), x - x_0] \\ + [A(x_0 + t(x - x_0)), h]\} \, dt$$

and

$$[A(x), h] = \int_0^1 \frac{d}{dt} [A(x_0 + t(x - x_0)), th] \, dt$$

$$= \int_0^1 \{[A'(x_0 + t(x - x_0)), x - x_0), th] \\ + [A(x_0 + t(x - x_0)), h]\} \, dt$$

$$= f'(x, h).$$

SECTION C. ROOTS AND EXTREMALS

C-1.   Theorem 1 (Hahn-Banach)

Every linear functional f defined on a subspace M of an NLS E can be extended to E without increasing its norm.

Proof

We shall prove the theorem for a separable space E. (See the remark below.) Let x_0 be a point of $E \sim M$ and let M_1 denote the linear hull of $M \cup \{x_0\}$. An arbitrary element y of M_1 is uniquely represented by $x + tx_0$, where $x \in M$ and $t \in R$. If we assume that this representation is not unique, we are led to the contradiction that x_0 is in M.

We now prove a useful inequality. For arbitrary z and z' in M, we have

$$f(z) - f(z') \leq \|f\| \, \|z + x_0 - z' - x_0\| \leq \|f\| \, \|z + x_0\| + \|f\| \, \|z' + x_0\|,$$

whence

$$f(z') + \|f\| \, \|z' + x_0\| \geq f(z) - \|f\| \, \|z + x_0\|.$$

Therefore, there exists a number c such that

$$\sup\{f(z) - \|f\| \, \|z + x_0\| : z \in M\} \leq c \leq \inf\{f(z') + \|f\| \, \|-z' - x_0\| : z' \in M\}.$$

Define the linear function f_1 on M_1 by $f_1(y) = f_1(x + tx_0) - f(x) - ct$. Clearly, f_1 is linear on M_1 and the restriction of f_1 to M is f. We now show that $\|f_1\| = \|f\|$. The definition of c implies that $|f(z) - c| \leq \|f\| \, \|z + x_0\|$ for all $z \in M$. For arbitrary $z \in M$ and $t \in R$, set $tz = x$; then

$$|f(x) - ct| \leq \|f\| \, \|x + tx_0\|,$$

showing that f has been extended to M_1 without increase of norm. (Examine the case $t < 0$ also.)

Since E is separable, it contains a countable dense set $x_1, x_2, \cdots$. We now construct the subspace M_{n+1} by taking the linear hull of M_n and $x_n \notin M_n$, and extend f_n to f_{n+1} in the way described above. We do this for all n and thus obtain a functional f_∞, which is defined on a set M_∞, which is dense in E, and which is such that $\|f_\infty\| = \|f\|$. For arbitrary $x \in E$, let $\{x_n\} \to x$ and set $F(x) = \lim_{n \to \infty} f_\infty(x_n)$. Clearly, F is linear. Since $|F(x)| = \lim_{n \to \infty} |f_\infty(x_n)| \leq \|f\| \, \|x\|$, F is bounded and $\|F\| \leq \|f\|$. Since the restriction of F to M is f, $\|F\| \geq \|f\|$; hence $\|F\| = \|f\|$.   Q.E.D.

Remark

For those familiar with set theory, we shall delete the hypothesis of separability. We well-order the set $E \sim M$, which is possible by the theorem of Zermelo. Suppose, then, that the elements of $E \sim M$ are arranged in a transfinite sequence $x_1, x_2, \cdots, x_\omega, x_{\omega+1}, \cdots$. As before, we extend the functional $f_1, f_2, \cdots, f_\omega, f_{\omega+1}, \cdots$. Because the norm is unchanged at each step, we arrive by transfinite induction to a functional F that extends f to all E.

E X E R C I S E

The above theorem can be generalized to the following theorem.

Theorem 2 (Hahn-Banach)

Let p be a functional on a linear space E such that $p(x + y) \leq p(x) + p(y)$, $p(\alpha x) = \alpha p(x)$, $\alpha \geq 0$ and x, y in E. Let f be a linear functional defined on a subspace M of E such that $f(x) \le p(x)$ for all $x \in M$. Then f can be extended to a linear functional F whose restriction to M is f and such that $F(x) \le p(x)$ for all $x \in E$.

Hint: Observe that if E were an NLS and f were bounded, we would have a p-like functional, namely, $x \to \|f\| \, \|x\|$. There is one difference: p is homogeneous only for nonnegative scalars α. Follow the above proof, replacing $\|f\| \, \|x\|$ by $p(x)$. You will eventually obtain

$$c - f(z) \le p(z + x_0),$$
$$c - f(z) \ge -p(-z - x_0).$$

Make the substitution $z = x/t$, using one of these for positive t and the other for negative t.

Write out the details of the extension of the subspace.

Corollary 1

Let M be a subspace of the NLS E and x_0 a point of E not in the closure of M. There exists a point $f \in E^*$ such that $f(x_0) = 1$ and $f(M) = 0$.

Proof

Let M_1 denote the linear hull of $M \cup \{x_0\}$. An arbitrary element z is uniquely represented as $y + tx_0$, where $y \in M$ and $t \in R$. Set $f'(z) = t$. Clearly, f' is linear on M_1. It is also bounded, for if $t \ne 0$,

$$\|z\| = \|y + tx_0\| = |t| \left\| \frac{y}{t} + x_0 \right\| \ge |t| \, d.$$

(If $-(y/t) = y' \in M$, then $\|y' - x_0\| = \|(y/t) + x_0\|$.) Thus, $|f'(z)| = |t| = \|z\|/d$. Therefore, since $\|f'\| \le 1/d$, $f' \in M_1^*$. To complete the proof, we extend f' to f in E^*.

EXERCISE

Show that $\|f\| = 1/d$. *Hint*: Take a sequence $\{y_n\}$ in M such that $\|x_0 - y_n\| \to d$ and use the fact that $f'(M) = 0$ and $f'(x_0) = 1$. Set $M = \{0\}$ and prove Corollary 2 below from Corollary 1.

Corollary 2

There exists a linear functional f on an NLS E such that, given $y \in E$, $f(y) = \|y\|$ and $\|f\| = 1$.

EXERCISE

Show that $\|x\| = \sup\{f(x) : \|f\| = 1\}$.

Corollary 3

If the conjugate E^* of an NLS E is separable, so is E.

Proof

Let $\{f_n\}$ be dense in E^*. Choose $x_n \in E$ so that

$$\|x_n\| \le 1 \quad \text{and} \quad f_n(x_n) \ge \frac{\|f_n\|}{2}.$$

The set M of all finite linear combinations of elements out of $\{x_n\}$ with rational coefficients is countable, and the closure $\overline{M}$ of M is a subspace. If $\overline{M}$ is not dense in E, there exists $x_0 \in E$ such that $\inf\{\|x_0 - x\| : x \in \overline{M}\} > 0$. By Corollary 1, there exists $f \in E^*$ such that $f(x_0) = 1$ and $f(M) = 0$. Take a sequence $\{f_{n_i}\}$ converging to f. Then

$$\|f_{n_i} - f\| \ge f_{n_i}(x_{n_i}) \ge \frac{\|f_{n_i}\|}{2}.$$

Since $\{f_{n_i}\} \to f$, it must be that $f = 0$, a contradiction.

Definition

Let K be a closed convex subset of an NLS with 0 an interior point of K. Let $I(x) = \{a : a > 0, a^{-1}x \in K\}$ and let $g(x) = \inf\{a : a \in I(x)\}$. We call g the *gauge function for K with respect to the origin.*

Lemma

The gauge function g satisfies $g(x) < 1$, $g(x) = 1$, or $g(x) > 1$, according as x is an interior point, boundary point, or exterior point of K, respectively.

Proof

Given $x \in E$, we have $I(x) = [p, \infty]$, where $p > 0$. Clearly, $g(x) = p^{-1}$. If $x \notin K$, $\|x\| > \|p^{-1}x\|$, showing that $g(x) > 1$. The remaining cases are proved similarly.

EXERCISE

Prove that:

(a) $g(x) \geq 0$.
(b) $g(ax) = ag(x)$ for $a \geq 0$.
(c) $g(x + y) \leq g(x) + g(y)$.

Hint: Take x and y in K and set $a = g(x)$, $b = g(y)$, $c = a + b$. Then

$$\frac{x + y}{c} = \frac{a(a^{-1}x) + b(b^{-1}y)}{a + b}.$$

(d)   What property must K possess for g to be a norm?

Lemma

Let K be a convex subset of an NLS with 0 an interior point of K. Assume $p \in E \sim \bar{K}$. There exists a nonzero continuous linear functional separating p and K.

Proof

Let g be the gauge function for K. Then $g(p) > 1$. Set $f_0(ap) = ag(p)$. Thus, f_0 is defined on the one-dimensional subspace of E, which consists of the multiples of p. Moreover, $f_0(ap) \leq g(ap)$, since if $a \geq 0$, $f_0(ap) = ag(p)$, while if $a < 0$, $f_0(ap) = af_0(p) < 0 \leq g(ap)$. By Theorem 2, f_0 can be extended to a linear functional f such that $f(x) \leq g(x)$ for all $x \in E$. It follows that $f(x) \leq 1$ for all $x \in K$, while $f(p) > 1$. Let N be an open sphere at the origin with radius γ such that $N \subset K$. If $\|h\| < \gamma$, $f(h) < f(p)$ and $f(-h) < f(p)$, so that $|f(h)| < f(p)$. Given $\varepsilon > 0$, set $\alpha = \varepsilon/f(p)$ and choose $\delta = \alpha\gamma$. Then, if $\|h\| < \gamma$, $\|\alpha h\| = \|h'\| < \delta$, whence $|f(h')| = \alpha f(h) < \varepsilon$. Since f is continuous at 0, it is continuous everywhere, by Section B-1.   Q.E.D.

Theorem 3

Let C be a closed convex subset of an NLS and let p be a point not in C. Then there exists a continuous linear functional f satisfying $f(x) \leq 1$ for all $x \in K$ while $f(p) > 1$.

Proof

Observe that the preceding lemma remains valid if the interior point of K is not the origin but some other point. We show that there exists an open convex set K' containing C such that $p \in E \sim \overline{K}'$. Let d denote the distance from p to C. (We do not claim that there exists a point on C where this distance is achieved.) For each $x \in C$, let S_x denote an open sphere of radius $d/2$. Let $K' = \cup \{S_x : x \in C\}$. Clearly, K' is open. It is also convex. Given y_1 and y_2 in K', set $y = ty_1 + (1 - t)y_2$. Let $x_i (i = 1,2)$ denote a point of C such that $y_i \in S_{x_i}$. Set $x = tx_1 + (1 - t)x_2$. Clearly, $x \in C$. Moreover,

$$\|y - x\| \le t\|y_1 - x_1\| + (1 - t)\|y_2 - x_2\| \le \frac{d}{2}.$$

Thus, $y \in S_x$, showing that $y \in K'$. Clearly, the distance from p to $K' = d/2$. It follows that there exists a continuous linear functional f such that $f(x) \le 1$ for all $x \in K'$ and $f(p) > 1$. Since $C \subset K'$, the theorem is proved.

EXERCISES

   1.   Closed convex subsets of an NLS are weakly closed.
   2.   If a convex set C has an interior point, then through every boundary point of C there is a supporting hyperplane.

C-2.   *Mean-Value Theorem*

We now generalize the generalized mean-value theorem of Section C-2, Chapter I, to normed linear spaces.

Theorem

Let f be a map between normed linear spaces E and F. Given x and k in E, assume that f is G-differentiable on the line segment joining x and $x + k$. Then

$$\|f(x + k) - f(x)\| \le \sup\left\{\frac{\|f'(x + tk, k)\|}{\|k\|} : t \in (0, 1)\right\}\|k\|.$$

Proof

Set $y = f(x + k) - f(x)$ and take $g \in F^*$ such that (see Section C-1, Corollary 2, p. 134) $g(y) = \|y\|$ and $\|g\| = 1$. Set $h(t) = g(f(x + tk))$. Thus,

$$h'(t) = \lim_{\Delta t \to 0} \frac{g(f(x + (t + \Delta t)k) - g(f(x + tk))}{\Delta t} = (gf)(x + tk, k).$$

Indeed, since f is actually F-differentiable on $\langle x, x + k \rangle$, we get $h'(t) =$

$g'(f(x + tk), f'(x + tk, k))$ by the chain rule, whereas since g is linear, $h'(t) = g(f'(x + tk, k))$. Thus,

$$g(y) = g(f(x + k) - f(x)) = h(1) - h(0) = h'(\theta), \qquad (0 < \theta < 1),$$

and $g(y) = g(f'(x + \theta k, k))$. On the other hand, since $g(y) = \|y\|$,

$$\|f(x + k) - f(x)\| = g(f'(x + \theta k, k)) = \left\| \frac{f'(x + \theta k, k)}{\|k\|} \right\| \cdot \|k\|;$$

whence the lemma.

E X E R C I S E S

1. Show that if f is a twice F-differentiable map between the NLS E and F, then the inequality

$$\|f(x + h) - f(x) - f'(x, h)\| \le \tfrac{1}{2} \sup_{\theta} \{\|f''(x + \theta h, \cdot, \cdot)\| 0 < \theta < 1\} \|h\|^2$$

is valid.

2. Show that if P is an operator from a convex subset E_0 of an NLS E to E_0, and P has a G-derivative at each point of E_0 and $\sup_{x \in E_0} \|P'(x)\| = a < 1$, then there exists a unique fixed point in E_0.

3. Assume that $f : E \to F$, where E and F are NLS. Assume that the G-derivative of f exists on every point of an open set E' of E. Let the mapping $x : h \to f'(x, h)$ be denoted by $f'(x)(\cdot)$. Assume that f is continuous at x_0 in E'. Prove that $f'(x)$ is the F-derivative of f at x_0.

C-3. *Reflexive Spaces, Locally Uniformly Convex Spaces, and Inverse Operators*

In what follows, a linear function f on a vector space E will be denoted by $f(\cdot)$, $[f, \cdot]$ or $[\cdot, f]$.

Definition

Let E be an NLS and E^{**} the conjugate of the B-space E^*. For $x \in E$, let $[\cdot, x]$ denote the linear map from E^* to R that assigns the value $[f, x]$ to each f in E^*. Let $C : E \to E^{**}$ denote the mapping that to each $x \in E$ assigns a linear function g on E^* such that $[g, f] = [f, x]$ for all $f \in E^*$. Since

$$\|g\| = \sup\{[f, x] : \|f\| = 1\} \le \|x\|, \, g \in E^{**}.$$

Moreover, since $\|x\| = \sup\{[f, x] : \|f\| = 1\}$ (Section C-1), and since it is readily verified that C is linear and 1-1, we see that C is an isometric isomorphism between E and the range of C. The map C is called the *canonical* or *natural embedding* from E to E^{**}.

Definition

A B-space E is called *reflexive* if the canonical embedding C from E to E^{**} is onto.

Theorem

A closed linear subspace in a reflexive B-space is a reflexive B-space.

Proof

Let M denote the subspace M^* and M^{**}, the conjugate of M and M^*, respectively. We take the following for typical elements in the various space we shall discuss: $x \in E$, $f \in E^*$, $g \in E^{**}$, $x' \in M$, $f' \in M^*$, and $g' \in M^{**}$. We write $f(x) = [f, x]$, $g(f) = [g, f]$, etc.

Let us denote by ξ the map that sends an element f in E^* to its restriction f' in M^*. Specifically, take $\xi(f)$ so that $[\xi(f), x'] = [f, x']$ for all $x' \in M$. Since $\|\xi(f)\| \le \|f\|$, $\xi(f)$ belongs to M^*. Moreover, $\xi(\cdot)$ is linear on E^*.

Since ξ is linear on E^* and g' is linear on M^*, the composition $g'\xi$ defined by $[g', \xi(f)] = g'\xi(f)$ for all $f \in E^*$ is linear on E^*. Moreover, $\|g'\xi\| \le \|g'\|$ so that $g'\xi \in E^{**}$. Set $\eta(g') = g'\xi$. Thus, if $f \in E^*$, $[g', \xi(f)] = [\eta(g'), f]$. Clearly, $\eta : M^{**} \to E^{**}$.

Now take g'_0 arbitrarily in M^{**} and set $g_0 = \eta(g'_0)$. Given $f' \in M^*$, let f be any extension of f' to E^*, so that $f' = \xi(f)$. For some $x_0 \in E$ and all $f' \in M^{**}$, we have

$$[g'_0, f'] = [g'_0, \xi(f)] = [\eta(g'_0), f] = [Cx_0, f] = [f, x_0],$$

where C is the canonical embedding. If $x_0 \in M$, then $[f, x_0] = [f', x_0]$ and we are done, since g'_0 was arbitrary in M^{**}.

Suppose, then, that $x_0 \notin M$. By Section C-1, Corollary 1, there exists $f \in E^*$ such that $[f, x_0] \ne 0$ and $[f, x'] = 0$ for all $x' \in M$. Since $[\xi(f), x'] = [f, x'] = 0$, $\xi(f) = 0 = [g'_0, \xi(f)] = [f, x_0]$, as above. This contradiction establishes that $x_0 \in M$, and proves the theorem.

Corollary

A B-space is reflexive if and only if its conjugate space is reflexive.

Proof

Let E be reflexive and let C be the canonical embedding of E into E^{**}. Let f be given in $(E^*)^{**} = (E^{**})^*$. Define $f \in E^*$ by $[f, x] = [f, Cx]$. Then $[g, f] = [Cx, f] = [f, x] = [f, Cx] = [f, g]$, showing that E^* is reflexive. Conversely, assume that E^* is reflexive. Then E^{**} is reflexive, and by the above theorem, the closed subspace $S = \{g = Cx : x \in E\}$ of E^{**} is reflexive. Since S is isometrically isomorphic to E, E is reflexive.

Theorem

Bounded subsets of a reflexive B-space E are weakly compact.

Proof

Let S be a bounded subset of E and let $\{x(j)\}$ be a sequence in S. Let M denote the linear hull of $\{x(j)\}$, namely, the set of all possible finite linear combinations of $\{x(j)\}$. Then $\overline{M}$, the closure of M, is a closed subspace of E. Let M_0 denote the set of all finite linear combinations of $\{x(j)\}$ with rational coefficients. M_0 is denumerable and dense in $\overline{M}$. Since $\overline{M}$ is a closed subspace, we have, by the above theorem, that $\overline{M}$ is a reflexive B-space. Since $\overline{M}$ is separable and reflexive, $\overline{M}^{**}$ is separable. By Section C-1, Corollary 3, $\overline{M}^*$ is separable. Let $\{f'_1, f'_2, \cdots\}$ be a dense subset of $\overline{M}^*$. Choose a subsequence $\{x(j(i, 1))\}$ such that $\{[f'_1, x(j(i, 1))]\}$ converges. Let $\{x(j(i, 2))]\}$ be a subsequence of $\{x(j(i, 1))\}$ such that $\{[f'_2, x(j(i, 2))]\}$ converges, and let $\{x(j(i, n))\}$ be a subsequence of $\{x(j(i, n - 1))\}$ such that $\{[f'_n, x(j(i, n))]\}$ converges. The sequence $\{x_i\} = \{x(j(i, i))\}$ has the property that $\{[f'_n, x_i]\}$ converges for every $n = 1, 2, 3, \cdots$ because, for $i \geq n$, $\{x(j(i, i))\}$ is a subsequence of $\{x(j(i, n))\}$. Thus,

$$\lim_{i \to \infty}[f'_n, x_i] = \lim_{i \to \infty}[g'_i, f'_n]$$

exists for each n. Since $\{f'_n : n = 1, 2, 3, \cdots\}$ is dense in $\overline{M}^*$ and because g'_i is uniformly bounded, we have by Section B-1 (the Banach-Steinhaus theorem) that there exists g' in $\overline{M}^{**}$ such that $\lim_i[f', x_i] = [g', f']$ for all f' in $\overline{M}^*$. Since $\overline{M}$ is reflexive, there is a point x' in $\overline{M}$ such that $[f', x_i] \to [f', x']$ for all f' in $\overline{M}^*$. Finally, because $x_i \in M$, for every $f \in E^*$ there corresponds an $f' \in \overline{M}^*$ such that $[f, x_i] = [f', x_i]$. Hence, $\{x_i\} \to x'$, which shows that S is weakly compact.

Corollary

Bounded and weakly closed subsets of Hilbert space are weakly compact.

Examples of reflexive spaces are Hilbert space and the space l_p (see Section B-1).

Definition

E is a locally uniformly convex B-space if at every point $z \in E$ such that $\|z\| = 1$, one has that $\|x_n\| = 1$ and $\|x_n + z\| \to 2$ imply $\|x_n - z\| \to 0$.

Theorem

If E is a locally uniformly convex B-space and $\{x_k\} \rightharpoonup y$ and $\{\|x_k\|\} \to \|y\|$, then $\{x_k\} \to y$.

Proof

Let

$$x'_k = \frac{x_k}{\|x_k\|} \quad \text{and} \quad y' = \frac{y}{\|y\|}.$$

Then $\{x'_k\} \rightharpoonup y'$ and $\|x'_k\| = \|y'\| = 1$. By a familiar consequence of the Hahn-Banach theorem, we may write

$$\lim \inf \|x'_k + y'\| = \lim \inf \sup\{f(x'_k + y') : \|f\| = 1\}$$

$$\geq \lim \inf f(x'_n + y') = 2f(y'),$$

provided $\|f\| = 1$. Thus,

$$\lim \inf \|x'_n + y'\| \geq 2 \sup\{f(y') : \|f\| = 1\} = 2.$$

Since $\|x'_k + y'\| \leq 2$, $\lim \sup \|x'_k + y'\| = 2$, and consequently

$$\lim \|x'_k + y'\| = 2.$$

Thus, $\{x'_k\} \to y'$; whence, since $\|x_k\| \to \|y\|$, $\{x_k\} \to y$.   Q.E.D.

EXERCISE

Show that Hilbert space is locally uniformly convex.

Definition

Let E and F be NLS and let A be a map from E to F. If, for each y in F, there exists a unique x in E such that $Ax = y$, then the map A is said to have an inverse. The mapping that assigns to each y in F the point x in E as its image is called the *inverse* of A and is denoted by A^{-1}. Thus, $x = A^{-1}y$.

EXERCISES

Assume A is linear and onto and prove the following:

1. If A^{-1} exists, it is linear.

2. A has an inverse iff $Ax = 0 \Rightarrow x = 0$.

3. A has a bounded inverse iff there exists a number $m > 0$ such that for all x in E, $\|Ax\| \geq m\|x\|$.

4. If $\|Ax\| \geq m\|x\|$, then $\|A^{-1}\| \leq 1/m$ and $\|A^{-1}y\| \geq \|A\|^{-1}\|y\|$.

Lemma

Let E be a reflexive B-space and $A \in B(E, E^*)$. If $[Ax, x] \geq m\|x\|^2$ for all $x \in E$ and some $m > 0$, then A has an inverse.

Proof

We must show that A is onto. If not, take[5] $f^0 \notin M = $ range A. Choose $g \in E^{**}$ such that $[g, f^0] = \inf\{\|f^0 - f\| : f \in M\}$, with $\|g\| = 1$, $g(M) = 0$. (See Corollary 1, Section C-1.) Take $\bar{x}$ in E so that $[g, f] = [f, \bar{x}]$ for all $f \in E^*$. Then $0 = [g, Ax] = [Ax, \bar{x}]$ for all $x \in E$. Thus, $[A\bar{x}, \bar{x}] = 0$ while $\|g\| = \|\bar{x}\| = 1$, a contradiction.

Lemma (Banach)

Let A be a linear operator mapping a B-space E into itself. Assume that $\|A\| < 1$. Then

(i) $(I - A)^{-1} = \sum\limits_{k=0}^{\infty} A^k$

(ii) $\|(I - A)^{-1}\| \leq (1 - \|A\|)^{-1}$.

Proof

Choose y arbitrarily in E and consider the mapping $x \to y + Ax$, which we denote by M. Clearly, M is a contraction in E and has a unique fixed point, which we denote by z. Thus, $z = y + Az$. Choose $x_0 = 0$ and consider the iterates $x_{n+1} = Mx_n$. We have $x_1 = y$, $x_2 = y + Ay, \cdots, x_n = y + Ay + \cdots + A^{n-1}y$. Thus,

$$z = \sum\limits_{k=0}^{\infty} A^k y = y + Az.$$

---

[5] Exercise: The range of A is closed.

Because the equation $y = (I - A)z$ has a unique solution z for every y in E, $(I - A)^{-1}$ exists, and consequently $z = (I - A)^{-1}y$. Therefore,

$$((I - A)^{-1})y = \sum_{k=0}^{\infty} A^k y,$$

proving (i). To prove (ii), we observe that

$$\|(I - A)^{-1}\| \le \sum_{k=0}^{\infty} \|A^k\| \le \sum_{k=0}^{\infty} \|A\|^k = (1 - \|A\|)^{-1}.$$

C-4. Newton's Method

We now turn our attention to a beautiful theorem due to L. V. Kantorovich, which generalizes the discussion of Chapter I, Section A-2. In the case of finding roots of real valued functions, the localization process consists of finding an interval on which the function changes sign. Kantorovich gives us conditions to check on a sphere in a Banach space, which will guarantee that an operator equation has a root on the sphere. Moreover, when these conditions obtain, the natural generalization of the Newton iteration of Chapter I, Section A-2, defines a sequence that converges to a root of the operator equation on the sphere. The essential condition is that the first approximation x_0 be sufficiently close to a root x^*. Thus, the number η_0 below tends to 0 if x_0 tends to x^*, and thus h_0 tends to 0. (We are assuming, of course, that the numbers β_0 and K remain bounded in a neighborhood of x^*.) Thus, in some neighborhood of x^*, $h_0 \le \frac{1}{2}$, and the theorem below will be in force.

In what follows, f is a nonlinear map between Banach spaces E and F. We denote the F-derivative $f'(x, \cdot)$ by $f'(x)$; the differential $f'(x, h)$ by $f'(x)h$, and the inverse operator $[f'(x)]^{-1}$ by $f'_{-1}(x)$.

Theorem 1 (Kantorovich)

Let a point x_0 in E be given for which $f'_{-1}(x_0)$ exists. Set

$$\eta_0 = \|f'_{-1}(x_0)f(x_0)\|$$

and let S denote the sphere $\{x \in E : \|x - x_0\| \le 2\eta_0\}$. Set $\beta_0 = \|f'_{-1}(x_0)\|$. If there exists a number K such that

$$\|f'(x) - f'(y)\| \le \frac{K\|x - y\|}{2} \qquad \text{for all } x \text{ and } y \text{ in } S$$

and such that $\beta_0\eta_0 K = h_0 \le \frac{1}{2}$, then f has a root in S and the Newtonian sequence $\{x_k\}$ defined by $x_{k+1} = x_k - f'_{-1}(x_k)f(x_k)$ converges to it. Furthermore, the rate of convergence of $\{x_k\}$ is quadratic.

Proof

We shall first establish that $\{x_k\}$ is well defined and is contained in S. Clearly, $x_1 \in S$, whence

$$\|f'(x_0) - f'(x_1)\| \leq \frac{K}{2} \|x_0 - x_1\| = \frac{K}{2} \eta_0$$

and

$$\|f'_{-1}(x_0)(f'(x_0) - f'(x_1))\| \leq \frac{\beta_0 K \eta_0}{2} = \frac{h_0}{2} \leq \frac{1}{4}.$$

By Banach's lemma, the operator $H = I - f'_{-1}(x_0)(f'(x_0) - f'(x_1))$ which is a bounded linear operator, from E to itself has an inverse, and

$$\|H^{-1}\| \leq \left(1 - \frac{h_0}{2}\right)^{-1}.$$

Also, $f'(x_0)H = f'(x_1)$ and $H^{-1}f'_{-1}(x_0) = f'_{-1}(x_1)$. Therefore,

$$\|f'_{-1}(x_1)\| \leq \|H^{-1}\| \|f'_{-1}(x_0)\| \leq \left(1 - \frac{h_0}{2}\right)^{-1} \beta_0 < (1 - h_0)^{-1} \beta_0 = \beta_1.$$

Now set $F_0(x) = x - f'_{-1}(x_0)f(x)$. Differentiating, we get $F'_0(x_0) = 0$. Also, $f'_{-1}(x_0)f(x_1) = x_1 - F_0(x_1)$ and $x_1 = x_0 - f'_{-1}(x_0)f(x_0) = F_0(x)$. Hence,

$$f'_{-1}(x_0)f(x_1) = F_0(x_0) - F_0(x_1)$$

$$= F_0(x_0) - F_0(x_1) - F'_0(x_0)(x_0 - x_1) = z.$$

Choose a functional g in E^* such that $g(z) = \|z\|$ and $\|g\| = 1$. By the mean value theorem (see Section C-2),

$$g(F_0(x_0) - F_0(x_1)) = (gF_0)'(\bar{x})(x_0 - x_1) \qquad where \; \bar{x} = \theta x_0 + (1 - \theta)x_1$$

for some θ in $(0, 1)$. Thus

$$g(z) = \|z\| = ((gF_0)'(\bar{x}) - gF'_0(x_0))(x_0 - x_1)$$

$$= g((F'_0(\bar{x}) - F'_0(x_0))(x_0 - x_1))$$

$$\leq \|g\| \|F'_0(\bar{x}) - F'_0(x_0)\| \|x_0 - x_1\|.$$

Since $F'_0(\bar{x}) - F'_0(x_0) = f'_{-1}(x_0)[f'(x_0) - f'(\bar{x}_1)]$,

$$\|F'_0(\bar{x}) - F'_0(x_0)\| \leq \frac{\beta_0 K \|x_1 - x_0\|}{2}.$$

Thus,

$$g(z) = \|f'_{-1}(x_0)f(x_1)\| \leq \frac{\beta_0 K \eta_0^2}{2} = \frac{h_0 \eta_0}{2}.$$

Since $f'_{-1}(x_1) = H^{-1}f'_{-1}(x_0)$, $f'_{-1}(x_1)f(x_1) = H^{-1}f'_{-1}(x_0)f(x_1)$ and therefore

$$\|f'_{-1}(x_1)f(x_1)\| \leq \left(1 - \frac{h_0}{2}\right)^{-1} \frac{h_0 \eta_0}{2} \leq \frac{1}{2} \frac{h_0 \eta_0}{1 - h_0} = \eta_1 \leq \eta_0.$$

Similarly, we define

$$h_1 = \beta_1 \eta_1 K = \frac{\beta_0}{1 - h_0} \cdot \frac{1}{2} \frac{h_0 \eta_0}{1 - h_0} \cdot K = \frac{1}{2} \frac{h_0^2}{(1 - h_0)^2} \leq 2h_0^2 \leq \frac{1}{2}.$$

Thus, in general, we set

$$\beta_n = \frac{\beta_{n-1}}{1 - h_{n-1}}, \quad \eta_n = \frac{1}{2} \frac{h_{n-1}\eta_{n-1}}{1 - h_{n-1}}, \quad h_n = \frac{\frac{1}{2}h_{n-1}^2}{(1 - h_{n-1})^2},$$

and $F_n(x) = x - f'_{-1}(x_n)f(x)$. Assume that for every $n = 1, 2, 3, \cdots, N$, the following inequalities make sense:

$$\|f'_{-1}(x_n)f(x_n)\| \leq \eta_n, \qquad \|f'_{-1}(x_n)\| \leq \beta_n, \qquad \text{and} \quad h_n = \beta_n \eta_n K \leq \frac{1}{2},$$

and that the point $x_n \in S$. Clearly, the point x_{N+1} is well defined. It is not immediately obvious, however, that $x_{N+1} \in S$. Because $x_n \in S$ and $h_n \leq \frac{1}{2}$ for ($1 \leq n \leq N$), we have (as in the above proof for $n = 1$) that $\eta_n \leq \frac{1}{2}\eta_{n-1}$ for ($1 \leq n \leq N$). It follows that

$$\|x_{N+1} - x_0\| \leq \|x_{N+1} - x_N\| + \|x_N - x_{N-1}\| + \cdots + \|x_1 - x_0\|$$
$$\leq \sum_{i=0}^{N} \eta_n < 2\eta_0,$$

showing that $x_{N+1} \in S$. As in the proof written out for $N = 0$, we find in turn that $h_{N+1} \leq \frac{1}{2}$, $\beta_{N+1} = (1 - h_N)^{-1}\beta_N$, and $\eta_{N+1} \leq \frac{1}{2}\eta_N$, which completes the induction.

We may similarly estimate that

$$\|x_{n+p} - x_n\| = \eta_n \sum_{j=0}^{p-1} 2^{-j} < 2\eta_n \qquad \text{for all } p = 1, 2, 3, \cdots.$$

Since $\{\eta_n\} \to 0$, it follows that the sequence $\{x_n\}$ is Cauchy. Because E is complete, $\{x_n\} \to x^*$ in E. To show that $f(x^*) = 0$, observe that

$$\|f'(x_n)\| \leq \|f'(x_0)\| + \|f'(x_n) - f'(x_0)\| \leq \|f'(x_0)\| + 2K\eta_0$$

and that $f'(x_n)(x_{n+1} - x_n) = f(x_n)$. It follows that $f(x_n) \to 0$ and $f(x^*) = 0$.
To estimate the rate of convergence, we observe that

$$h_1 = \frac{1}{2}\left(\frac{h_0}{1 - h_0}\right)^2 < 2h_0^2; \quad h_2 \le 2h_1^2 \le 8h_0^4, \cdots, \quad h_n \le \frac{1}{2}(2h_0)^{2^n}.$$

Since $\eta_n = \frac{1}{2}h_{n-1}(1 - h_{n-1})^{-1}\eta_{n-1}$

$$\le h_{n-1}\eta_{n-1} \le h_{n-1}(h_{n-2}\eta_{n-2})$$

$$\le h_{n-1}(h_{n-2}(h_{n-3}\eta_{n-3}))$$

$$\le \cdots \le \frac{1}{2^n}(2h_0)^{2^{n-1}}(2h_0)^{2^{n-2}} \cdots (2h_0)\eta_0$$

$$= 2^{-n}(2h_0)^{2^n - 1}\eta_0.$$

(Observe that the sum $\sum_{k=1}^{n} 2^{n-k}$ as a binary expansion is all ones.) Thus,

$$\|x^* - x_n\| \le 2^{-n+1}(2h_0)^{2^n - 1}\eta_0.$$

Remark 1

Suppose that f is twice G-differentiable on S and

$$M = \sup\{\|f''(x, h, h)\| : x \in S, h \in E, \text{ and } \|h\| = 1\} < \infty.$$

Then, if $h'_0 = \beta_0\eta_0 M$, and $2h'_0 \le h_0$, then f has a root in S to which $\{x_k\}$ converges quadratically whenever $h'_0 \le \frac{1}{2}$. Thus, if M is available, it is advantageous to use it rather than K.

Proof

We have that $\|f'(x) - f'(y)\| \le M\|x - y\|$ by the generalized mean-value theorem. Hence, $M \le K/2$. The remark follows by use of the inequality $\|f(x + h) - f(x) - f'(x, h)\| \le \frac{1}{2}M\|h\|^2$. (See Section C-2, Exercise 1.)

Exercise

Sketch in the details of the above proof.

Remark 2

The sphere S can be replaced by a somewhat smaller sphere, namely, a sphere with center x_0 and radius $\eta_0 N(h_0)$, where

$$N(h) = \frac{1 - \sqrt{1 - 2h}}{h}.$$

Proof

We need the following identity:

(I) $$\eta_n = \eta_n N(h_n) - \eta_{n+1} N(h_{n+1}).$$

We calculate:

$$\eta_{n+1} N(h_{n+1}) = \eta_{n+1} \frac{1 - \sqrt{1 - h_{n+1}}}{h_{n+1}}$$

$$= \frac{1}{2} \frac{h_n \eta_n}{1 - h_n} \cdot \frac{1 - \sqrt{1 - (h_n^2/(1 - h_n)^2)}}{\frac{1}{2}(h_n^2/(1 - h_n)^2)}$$

$$= \eta_n \frac{1 - h_n - \sqrt{1 - 2h_n}}{h_n}$$

$$= \eta_n N(h_n) - \eta_n,$$

where we have used h_n and η_n as defined in the proof of the above theorem. By the triangle inequality,

$$\|x_0 - x_n\| \le \eta_0 + \eta_1 + \cdots + \eta_{n-1} = \eta_0 N(h_0) - \eta_n N(h_n) \le \eta_0 N(h_0).$$

Q.E.D.

Theorem 3 (Kantorovich)

Assume now that the sphere S of Theorem 1 is replaced by an open sphere S' with center x_0 and radius $\eta_0 L(h_0)$, where $L(h) = (1 + \sqrt{1 - 2h})/h$. Then the equation $f(x) = 0$ has a unique root in this sphere.

Proof

The same computation as in the above remark validates the identity

$$\eta_n = \eta_n L(h_n) - \eta_{n+1} L(h_{n+1}).$$

This identity implies that

$$L^2(h_n) = \frac{2\eta_{n+1} L(h_{n+1})}{h_n \eta_n}.$$

To verify this, calculate that

$$L^2(h_n) = \left(\frac{1 + \sqrt{1 - 2h_n}}{h_n} \right)^2 = \frac{2(1 + \sqrt{1 - 2h_n} - h_n)}{h_n^2}$$

$$= \frac{2}{h_n}[L(h_n) - 1] = \frac{2\eta_{n+1}}{h_n \eta_n} L(h_{n+1})$$

by the above identity. Assume $\tilde{x}$ is a root of f in S'. Since $\tilde{x} \in S'$, $\|\tilde{x} - x_0\| < \eta_0 L(h_0)$. Thus, for some θ, $0 \leq \theta < 1$, $\|\tilde{x} - x_0\| = \theta \eta_0 L(h_0)$. We may also observe that $F_0(\tilde{x}) = \tilde{x}$. Arguing as in the proof of Theorem 1, we have

$$\|\tilde{x} - x_1\| = \|F_0(\tilde{x}) - F_0(x_0)\|$$
$$= \|F_0(\tilde{x}) - F_0(x_0) - F'_0(x_0)(x - x_0)\| \leq \tfrac{1}{2}\beta_0 K \|\tilde{x} - x_0\|^2$$
$$= \tfrac{1}{2}\beta_0 K \theta^2 L^2(h_0)\eta_0^2 = \theta^2 L(h_1)\eta_1.$$

Thus, $\|\tilde{x} - x_0\| = \theta L(h_0)\eta_0$, while $\|\tilde{x} - x_1\| = \theta^2 L(h_1)\eta_1$. Again by induction, $\|\tilde{x} - x_n\| = \theta^{2^n} L(h_n)\eta_n$.

From the definition of $L(h_n)$,

$$L(h_n)\eta_n \leq \frac{2\eta_n}{h_n} = \frac{2}{\beta_n K} < \frac{2}{\beta_0 K} \, ;$$

thus, $\{x_n\} \to x$. Since $\{x_n\} \to x^*$, $\tilde{x} = x^*$, showing that x^* is unique.

Lemma

Let x^*, x_1, x_0, $N(h)$, η_0 and $F_0(x)$ be defined as above, and suppose that the conditions of Theorem 1 are satisfied, with S being the sphere

$$\{x: \|x - x_0\| \leq \eta_0 N(h_0)\}.$$

If a point $x \in E$ satisfies $\|x - x^*\| \leq \|x_1 - x^*\|$ and $\|x - x_0\| \leq N(h_0)\eta_0$, then $x' = F_0(x)$ satisfies

$$\|x' - x^*\| \leq \frac{h_0 N(h_0)}{2} \|x - x^*\| \quad \text{and} \quad \|x' - x_0\| \leq N(h_0)\eta_0 .$$

Proof

Since $F'_0(x_0) = 0$,

$$\|x' - x^*\| = \|F_0(x) - F_0(x^*)\|$$
$$\leq \sup\{\|F'_0(\tilde{x})\|\|x - x^*\| : \tilde{x} = x + \theta(x^* - x) \quad \text{and} \quad 0 < \theta < 1\}$$
$$= \sup\{\|F'_0(\tilde{x}) - F'_0(x_0)\|\|x - x^*\|\}$$
$$= \sup\{\|f'_{-1}(x_0)[f'(x_0) - f'(\tilde{x})]\|\|x - x^*\|\}$$
$$\leq \beta_0 \frac{K}{2} \|x - x^*\| \sup\{\|x_0 - \tilde{x}\|\}.$$

Since

$$\|\bar{x} - x_0\| = \|(1 - \theta)x + \theta x^* - (1 - \theta)x_0 - \theta x_0\|$$

$$\leq (1 - \theta)\|x - x_0\| + \theta\|x^* - x_0\|$$

$$\leq \max\{\|x - x_0\|, \|x^* - x_0\|\} \leq N(h_0)\eta_0,$$

it follows that

$$\|x' - x^*\| \leq \beta_0 \frac{K}{2} N(h_0)\eta_0\|x - x^*\| = h_0 N(h_0) \frac{\|x - x^*\|}{2}.$$

For the second half, we have

$$\|x' - x_0\| \leq \|x' - x_1\| + \|x_1 - x_0\|$$

$$= \|F_0(x) - F_0(x_0) - F'_0(x_0)(x - x_0)\| + \eta_0,$$

and arguing as in the proof of Theorem 1,

$$\|x' - x_0\| \leq \tfrac{1}{2}\beta_0 K\|x - x_0\|^2 + \eta_0 \leq \tfrac{1}{2}\beta_0 K[N(h_0)\eta_0]^2 + \eta_0.$$

Observe now that N satisfies the same identity that L does (see proof of Theorem 2), namely: $N^2(h_0) = 2[N(h_0) - 1]h_0^{-1}$, whence $\tfrac{1}{2}\beta_0 K_0^2 N^2(h_0) = \eta_0 N(h_0) - \eta_0$, which completes the proof.

Theorem 4 (Kantorovich)

Assume the hypotheses of the above lemma and define the iteration $x'_{n+1} = F_0(x'_n)$. Then $\{x'_n\} \to x^*$ at the rate of a geometric progression.

Proof

Set $x = x_1$. Clearly, x satisfies the conditions of the above lemma, so that $F_0(x) = F_0(x_1) = x'$ satisfies

$$\|x' - x^*\| \leq h_0 N(h_0) \frac{\|x - x^*\|}{2} \quad \text{and} \quad \|x' - x_0\| \leq N(h_0)\eta_0.$$

Thus, x' satisfies the conditions of the above lemma, since $h_0 N(h_0) \leq 1$. Set $x_1 = x'_1$ and $x' = x'_2$. Since x'_2 satisfies the conditions of the above lemma, then so does x'_3, and so forth. Clearly,

$$\|x'_{n+1} - x^*\| \leq \left(\frac{h_0 N(h_0)}{2}\right)^n \|x_1 - x^*\| \leq \left(\frac{1}{2}\right)^n \|x_1 - x^*\|.$$

Q.E.D.

Remark

Thus, if we are willing to sacrifice speed of convergence, the operator $f'(x)$ need be inverted only once.

C-5. Minimizing Functionals on NLS

Our object in this section is to generalize the results of Sections D-3 and D-4, Chapter I. In what follows, if E is an NLS, $g \in E^*$ and x in E, then we write $[g, x]$ for $g(x)$. Recall that if E and F are normed linear spaces (see Section C-3, p. 142), A is a bounded linear operator from E to F; in short, $A \in B(E, F)$, and if A is onto, then A^{-1} exists and belongs to $B(F, E)$, if and only if for some $m > 0$ and all $x \in E$, $\|Ax\| \geq m\|x\|$; and furthermore, that $m\|x\| \leq \|Ax\| \leq M\|x\|$ for all x in E implies that $M^{-1}\|y\| \leq \|A^{-1}y\| \leq m^{-1}\|y\|$ for all $y \in F$.

Let ϕ denote a bounded map from S to E satisfying two conditions: that $[f'(x), \phi(x)] \geq 0$ and, given $\varepsilon > 0$, that there exists $\delta > 0$ such that $[f'(x), \phi(x)] < \delta$ implies $\|f'(x)\| < \varepsilon$. Some examples of such mappings are the following:

(i)   Let $A \in B(E^*, E)$ such that $[y, Ay] \geq \sigma\|y\|^2$ for all $y \in E^*$ and some $\sigma > 0$. Let $\phi(x) = Af'(x)$ and choose $\delta = \varepsilon^2\sigma$. Then $\|f'(x)\| < \varepsilon$. As a possible candidate for the operator A, suppose f is twice F-differentiable on E. Assume that for some $\mu > 0$ and some x in S, the operator $f''(x)$ in $B(E, E^*)$ is onto[6] and "bounded below"; that is, $[f''(x)z, z] \geq \mu\|z\|^2$ for all z in E. Then $\|f''(x)z\| \geq \mu\|z\|$, showing that $f''(x)$ has an inverse $[f''(x)]^{-1} = A \in B(E^*, E)$. Since A has a bounded inverse, there exists a number $\sigma > 0$ such that $\|Ay\| \geq \sigma\|y\|$ for all $y \in E^*$. Set $z = Ay$. Then $[f''(x)z, z] = [y, Ay] \geq \mu\sigma^2\|y\|^2$, showing the candidacy of A.

(ii)   Suppose E is a reflexive Banach space. By the weak compactness of the unit sphere in E and the weak continuity of $f'(x)$, it follows that for some z_0, $\|z_0\| = 1$, $[f'(x), z_0] = \|f'(x)\|$. Set $\phi(x) = z_0\|f'(x)\|$. Because $[f'(x), \phi(x)] = \|f'(x)\|^2$, $\phi(x)$ is the analog of the gradient in Hilbert space.

(iii)   Since $\|f'(x)\| = \sup\{[f'(x), z]: \|z\| = 1\}$, if $0 < \alpha < 1$, a point z_0 exists such that $[f'(x), z_0] \geq \alpha\|f'(x)\|$. If for fixed α and all $x \in S$, we can find z_0, we may take $\phi(x) = z_0\|f'(x)\|$.

In what follows, let

$$\Delta(x, \gamma) = f(x) - f(x - \gamma\phi(x)) \quad \text{and} \quad g(x, \gamma) = \frac{\Delta(x, \gamma)}{\gamma[f'(x), \phi(x)]}.$$

---

[6] Recall that by the last lemma of Section C-3 the onto property follows automatically when the space E is a reflexive B-space.

Assume that E is a normed linear space and that S is the level set of f at x_0 in E. Select σ such that $0 < \sigma < \frac{1}{2}$.

Theorem 1

Assume that on S the F-derivative f' exists and is uniformly continuous. Set $x_{k+1} = x_k$, when $[f'(x_k), \phi(x_k)] = 0$; otherwise, choose[7] γ_k so that $\sigma < g(x_k, \gamma_k) \leq 1 - \sigma$ when $g(x_k, 1) < \sigma$ or $\gamma_k = 1$ when $g(x_k, 1) \geq \sigma$, and set $x_{k+1} = x_k - \gamma_k \phi(x_k)$.

(a) If S is bounded or f is bounded below, then $\{f'(x_k)\}$ converges to 0 while $\{f(x_k)\}$ converges downward to a limit L. If S is compact, then every cluster point of $\{x_k\}$ is a zero of f'. Moreover, if $\{\phi(x_k)\} \to 0$ and f' has finitely many zeros, $\{x_k\}$ converges.

(b) If S is convex and bounded and f is convex, $L = \inf\{f(x): x \in S\} = \theta$. If, in addition, E is a reflexive B-space, then every weak cluster point of $\{x_k\}$ minimizes f on E. If E is a uniformly convex B-space and $f(x) = \|x\|$, then $\{x_k\}$ converges.

(c) Assume that the Gateaux derivative f'' exists on S and satisfies

$$\mu\|z\|^2 \leq [f''(x)z, z] \leq M\|z\|^2$$

for all $x \in S$, $z \in E$ and some $\mu > 0$. Assume that S is convex and E is complete. Then $\{x_k\}$ converges to a unique minimizer of f on E.

Proof

(a) Assume $[f'(x), \phi(x)] \neq 0$. We have that

$$\Delta(x, \gamma) = [f'(\xi), \phi(x)] = \gamma[f'(x), \phi(x)] + \gamma[f'(\xi) - f'(x), \phi(x)],$$

where ξ lies " between " x and $x - \gamma(x)$. Thus,

$$g(x, \gamma) = 1 + \frac{[f'(\xi) - f'(x), \phi(x)]}{[f'(x), \phi(x)]}.$$

Since f' is continuous at x and $\|\xi - x\| \leq \gamma\|\phi(x)\|$, $g(x, 0) = 1$. If $g(x, 1) < \sigma$, then g, being continuous, takes on all values between 1 and σ; there exists, therefore, numbers $\gamma > 0$ so that $\sigma \leq g(x, \gamma) \leq 1 - \sigma$. Thus,

$$\Delta(\gamma_k, x_k) = \gamma_k g(\gamma_k, x)[f'(x_k), \phi(x_k)] \geq \gamma_k \delta[f'(x_k), \phi(x_k)] > 0, \, f(x_{k+1}) < f(x_k)$$

and $x_{k+1} \in S$. Assume that $\{[f'(x_k)\phi(x_k)]\} \nrightarrow 0$. There exists, then, a subsequence $\{x_k\}$ and a number $\varepsilon > 0$ such that $[f'(x_k), \phi(x_k)] \geq \varepsilon$. It follows,

---

[7] If the Gateaux differential f'' satisfies $f''(x, h, h) \leq \|h\|^2/\gamma_0$ for all h in E, x in S, and some $\gamma_0 > 0$, we can choose γ_k to satisfy $\delta \leq \gamma_k \leq 2\gamma_0 - \delta$ with $0 < \delta \leq \gamma_0$. The method of steepest descent could also be employed.

moreover, that $\{\gamma_k\}$ is bounded away from 0. If not, take a thinner subsequence $\{\gamma_k\}$, if necessary, such that $\{\gamma_k\} \to 0$. Since $\phi(x_k)$ is bounded on S, it follows by the uniform continuity of f' on S that

$$\{[f'(\xi_k) - f'(x_k), \phi(x_k)]\} \to 0,$$

and therefore $\{g(x_k, \gamma_k)\} \to 1$, contradicting that $g(x_k, \gamma_k) \leq 1 - \sigma$ for all k. There exists, therefore, a number $q > 0$ such that $\gamma_k \geq q$. Hence, $\Delta(x_k, \gamma_k) \geq q\, \delta\varepsilon$, from which we may contradict the hypothesis that f is bounded below. Thus, $\{[f'(x_n), \phi(x_k)]\} \to 0$. The remainder of the proof is as Chapter I, Section D-3, after we observe that S bounded implies that f is bounded below on S. (See exercise below.)

(b)   Given $\varepsilon > 0$, choose $z' \in E$ such that $f(z') \leq \theta + \varepsilon/2$. Because f' exists at x_k and f is convex, $f(z') \geq f(x_k) + [f'(x_k), z' - x_k]$. Since $\{f'(x_k)\} \to 0$ and S is bounded for all k sufficiently large, $f(x_k) \leq f(z') + \varepsilon/2 = \theta + \varepsilon$, showing that $L = \theta$.

By Section C-3, if S is convex, closed, and bounded, then S is weakly compact. Since f is convex, the sets $\{x \in E: f(x) \leq k\}$ are closed, convex, and weakly closed, for all k. Thus, f is weakly lower semicontinuous.

If z is a weak cluster point of $\{x_k\}$, then for an appropriate subsequence, $\lim \inf f(x_k) = L \geq f(z)$.

(c)   The hypotheses of (c) and the generalized mean-value theorem of Section C-4 imply that f' is Lipschitz continuous. Moreover, the set S is bounded. Otherwise, S would contain an unbounded sequence, say $\{z_k\}$. By Taylor's theorem, if $u \in S$,

$$f(z_k) \geq f(u) + \|z_k - u\|\left[(\|z_k - u\|)\frac{\mu}{2} - \|f'(u)\|\right],$$

showing that $f(z_k) \geq f(x_0)$ for large k, whence S must be bounded. We now show that the sequence $\{x_k\}$ is Cauchy. Again by Taylor's theorem, if $s > k$,

$$f(x_s) - f(x_k) \geq [f'(x_k), x_s - x_k] + \frac{\mu\|x_s - x_k\|^2}{2}.$$

Since S is bounded, $\|x_s - x_k\| \leq D$, where D is the diameter of S. Thus,

$$\|x_s - x_k\|^2 \leq \frac{2}{\mu}\{f(x_s) - f(x_k) + D\|f'(x_k)\|\},$$

which shows that $\{x_s\}$ is a Cauchy sequence. By the completeness of E, $\{x_s\}$ has a limit, say z, in E, and $f'(z) = 0$. If z is not unique, then $f'(z_1) = f'(z_2) = 0$, $z_1 \neq z$. Thus,

$$f(z_1) - f(z_2) \geq \frac{\mu}{2}\|z_1 - z_2\|^2 \leq f(z_2) - f(z_1),$$

a contradiction. Hence, z is unique and is a minimizer of f.

E X E R C I S E

Prove that S bounded and f' uniformly continuous imply that f' is bounded on S. If f' is bounded on S, show that f is bounded below on S, when S is bounded.

We now consider the possibility of accelerating as we did in Chapter I, Section D-4, by means of Newton steps.

Suppose that at the given point x_0 the function f satisfies the conditions of (i), p. 150, for example $\phi(x) = f''_{-1}(x_0)f'(x)$, where $f''_{-1}(x_0) = [f''(x_0)]^{-1}$. The corresponding iteration is $x_{n+1} = x_n - \gamma_n f''_{-1}(x_0)f'(x_n)$. This algorithm, when $\gamma_n \equiv 1$, is known as the "modified" Newton's method (see Theorem 4, Section C-4 above). In a similar manner, if f''_{-1} exists and is "uniformly bounded below" on S, we may define $\phi(x_n) = f''_{-1}(x_n)f'(x_n)$. We shall do this below. It is clear from what has already been said that ϕ satisfies hypotheses of the above theorem. Our object now is to formulate an algorithm, using $f''_{-1}(x_n)f'(x_n) = \phi(x_n)$, that will converge at a superlinear rate.

In the following theorem, we set

$$\Delta(x, \gamma) = f(x) - f(x - \gamma f''_{-1}(x)f'(x))$$

and

$$g(x, \gamma) = \frac{\Delta(x, \gamma)}{\gamma[f''_{-1}(x)f'(x), f'(x)]}.$$

Theorem 2

Assume that E is a B-space and the level set $S = \{x \in E : f(x) \leq f(x_0)\}$ is convex. For each x in S assume that the F-derivative f'' is continuous on S, $f''(x)$ is onto, $\|f''(x)\| \leq M$, and $[f''(x)z, z] \geq m\|z\|^2$ for some $m > 0$ and all z in E. Set $x_{k+1} = x_k - \gamma_k f''_{-1}(x_k)f'(x_k)$, where γ_k is chosen so that for $\theta < \frac{1}{2}, 0 < \theta \leq g(x_k, \gamma_k) \leq 1 - \theta$, with $\gamma_k = 1$ if possible. Then:

(a) There exists a number N such that if $k > N$, then $\gamma_k = 1$.
(b) There is a unique minimizer of f and the sequence $\{x_k\}$ converges to it faster than any geometric progression.

Proof

We have for all x in S that $M\|z\|^2 \geq [f''(x)z, z] \geq m\|z\|^2$ and $m^{-1}\|y\|^2 \geq [y, f''_{-1}(x)y] \geq mM^{-2}\|y\|^2$, since $[f''(x)z, z] = [f''(x)f''_{-1}(x)y, f''_{-1}(x)y] = [y, f''_{-1}(x)y] \geq mM^{-2}\|y\|^2$. Thus, if $\phi(x) = f''_{-1}(x)f'(x)$, then

$$[f'(x), \phi(x)] \geq mM^{-2}\|f'(x)\|^2,$$

showing that ϕ satisfies the conditions of Theorem 2. Since f'' is bounded on S, f' is Lipschitz continuous, by the generalized mean-value theorem (Section C-4). By (c) of Theorem 1, $\{x_k\}$ converges to a unique minimizer of f.

Expand $\Delta(x, \gamma)$ to two terms in the Taylor series with the remainder $[f''(\xi)h, h]$, where $h = \gamma f''_{-1}(x)f'(x)$. Set $f''(\xi) = f''(x) + f''(\xi) - f''(x)$. Then

$$g(x, \gamma) = 1 - \frac{\gamma}{2} - \frac{\gamma[(f''(\xi) - f''(x))f''_{-1}(x)f'(x), f''_{-1}(x)f'(x)]}{2[f'(x), f''_{-1}(x)f'(x)]}$$

$$\geq 1 - \frac{\gamma}{2} - \gamma\|f''(\xi) - f''(x)\| \frac{M^2}{2m^3}.$$

Thus,

$$\left| g(x, \gamma) - 1 + \frac{\gamma}{2} \right| \leq \gamma\|f''(\xi) - f''(x)\| \frac{M^2}{2m}.$$

Since $\xi(\gamma_k)$ lies between x_k and $x_{k+1}, x_0, \xi_0, x_1, \cdots$ is a Cauchy sequence; and it, together with its limit z, is a compactum. Consequently, on this compactum, f'' is uniformly continuous, so that $\{\|f''(\xi(\gamma_k)) - f''(x)\|\}$ converges to 0, showing that the choice $\gamma_k = 1$ is eventually feasible.

To prove (b) we write

$$x_{k+1} - z = x_k - z - \gamma_k f''_{-1}(x_k)f'(x_k) = x_k - z - \gamma_k f''_{-1}(x_k)f''(x_k)(x_k - z)$$

$$+ \gamma_k f''_{-1}(x_k)[f''(x_k)(x_k - z) - f'(x_k)].$$

Thus,

$$\|x_{k+1} - z\| = \|x_k - z - \gamma_k(x_k - z)\| + \gamma_k\|f''_{-1}(x_k)\|\|f''(x_k)(x_k - z) - f'(x_k)\|.$$

Since f' is F-differentiable at x_k,

$$\|f'(z) - f'(x_k) - f''(x_k)(z - x_k)\| < \varepsilon\|z - x_k\|.$$

Thus,

$$\|x_{k+1} - z\| = (1 - \gamma_k)\|x_k - z\| + \gamma_k m^{-1}\varepsilon\|z - x_k\|.$$

Q.E.D.

Remarks

(1) Both sides of the inverse of f''_{-1} are used in the proof.

(2) Under the hypothesis of Theorem 2, the analog of the modified Newton process, namely, choosing $\phi(x) = f''_{-1}(x_0)f'(x)$ or $f''_{-1}(x_k)f'(x)$, with k fixed, also will generate a sequence converging to a unique minimizer of f. Since

$$\|x_{k+1} - z\| = \|x_k - z - \gamma_k f''_{-1}(x_0)f''(z)(x_k - z)\|$$

$$+ \gamma_k\|[f''_{-1}(x_0)]\| \varepsilon \|x_k - z\|m^{-1}$$

when $\|x_k - z\| < \delta$, the rate of convergence is eventually geometric, provided $\|I - \gamma_k f''_{-1}(x_0) - f''(z)\| < 1$. Since

$$\|I - \gamma_k f''_{-1}(x_0) f''(z)\| \leq 1 - \gamma_k + \gamma_k \|f''_{-1}(x_0)\| \|(f''(x_0) - f''(z))\|,$$

if $\|f''(x_0) - f''(z)\|$ is sufficiently small, $\gamma_k \equiv 1$ will generate a sequence converging to z at the rate of geometric progression. A sufficient condition for the global geometric convergence would be $(M/m) < \frac{1}{2}$, since $\|f''(x)\| \leq M$ and $\|f''_{-1}(x)\| \leq m^{-1}$.

SECTION D. APPLICATIONS TO INTEGRAL EQUATIONS

D-1. Resolvent Kernel

Let C belong to $B(E, E)$, and suppose that C is close to identity operator E in the sense that $\|I - C\| < 1$. If $A = I - C$ and $\|A\| < 1$, then $C = I - A$ has an inverse by Banach's lemma (Section C-3). Thus, given $y \in E$, the equation $y = Cx$ has a unique solution.

Clearly, the equation

$$y = (I - \lambda A)x \qquad \text{(I)}$$

also has a unique solution if $\lambda \|A\| < 1$. Indeed, by Section C-3 we have that

$$(I - \lambda A)^{-1} = I + \lambda A + (\lambda A)^2 \cdots.$$

As an application of Equation (I), we consider a "Fredholm integral equation." This equation arises in "Sturm-Liouville" problems for differential equations. Given x and y in $C[0, 1]$ and k continuous on the square $\{(s, t): 0 \leq s \leq 1, 0 \leq t \leq 1\}$, set

$$y(s) = x(s) - \lambda \int_0^1 k(s, t)x(t)\, dt. \qquad \text{(II)}$$

Let

$$(I - \lambda A)x = x(\cdot) - \lambda \int_0^1 k(\cdot, t)x(t)\, dt.$$

Then $(I - \lambda A)$ is an element of $B(E, E)$ with $E = C[0, 1]$. The boundedness of $(I - \lambda A)$ follows because

$$\|A\| \leq \sup_s \int_0^1 |k(s, t)|\, dt.$$

Clearly, for some positive λ, $\lambda \|A\| < 1$, showing that for such λ, Equation (II) has a unique solution $x \in C[0, 1]$.

Since

$$(I - \lambda A)^{-1} = \left(I + \sum_{k=1}^{\infty} (\lambda A)^k\right) = I + \lambda B,$$

we have

$$(I - \lambda A)(I + \lambda B) = I + \lambda B - \lambda A(I + \lambda B) = I,$$

and

$$(I + \lambda B)(I - \lambda A) = I + \lambda B - \lambda(I + \lambda B)A = I;$$

whence

$$B = A(I + \lambda B) \tag{IIIa}$$

$$B = (I + \lambda B)A \tag{IIIb}$$

The operator

$$B = \lambda^{-1} \sum_{k=1}^{\infty} (\lambda A)^k$$

is called the *resolvent kernel*.

We observe that $\lambda By(s)$ has the value

$$\int_0^1 [\lambda k(s, t) + \lambda^2 \int_0^1 k(s, t_1)k(t_1, t)\, dt_1 + \cdots$$

$$+ \lambda^n \int_0^1 dt_n \int_0^1 dt_{n-1} \cdots \int_0^1 k(s, t_{n-1})k(t_n, t_{n-2}) \cdots k(t_1, t)\, dt_1 + \cdots]y(t)\, dt$$

which is of the form

$$\lambda By(s) = \lambda \int_0^1 r(s, t)y(t)\, dt.$$

Therefore, we set

$$By(s) = \int_0^1 r(s, t)y(t)\, dt.$$

By Equation (IIIa), recalling that

$$Ax = \int_0^1 k(\cdot, t)x(t)\, dt,$$

we have

$$\int_0^1 r(s, t)x(t)\, dt = \int_0^1 k(s, u)\left[x(u) + \lambda \int_0^1 r(u, t)x(t)\, dt\right] du$$

$$= \int_0^1 k(s, t)x(t)\, dt + \lambda \int_0^1 x(t)\left[\int_0^1 k(s, u)r(u, t)\, du\right] dt.$$

Since this equation is satisfied for every x in $C[0, 1]$, we get Equation (IVa) below. (Why?) The Equation (IVb) is a consequence of Equation (IIIb) by similar reasoning. Thus,

$$r(s, t) = k(s, t) + \lambda \int_0^1 k(s, u) r(u, t) \, du, \tag{IVa}$$

$$r(s, t) = k(s, t) + \lambda \int_0^1 r(s, u) k(t, u) \, du. \tag{IVb}$$

The Equations (IV) are called the integral equations for the resolvent kernel.

Equation (II) is called an equation of the second kind. An equation of the first kind is

$$y(s) = \lambda \int_0^1 k(s, t) x(t) \, dt.$$

D-2. *Solution by Gradient Method*

Our object now is to employ the techniques of Section C-5 to obtain another way of solving Equation (II). To this end we define an inner product in the space $C[0, 1]$ as in Section B-2; that is, for f and g in $C[0, 1]$, set

$$[f, g] = \int_0^1 f(t) g(t) \, dt.$$

Under the norm $[\cdot, \cdot]^{1/2}$, $C[0, 1]$ is an NLS. Let us further assume that k is symmetric, that is, $k(s, t) = k(t, s)$. It is readily verified that if

$$Ax = \int_0^1 k(\cdot, t) x(t) \, dt,$$

then $[Ax, y] = [x, Ay]$. Let $B = I - \lambda A$ and set $f(x) = [x, Bx] - 2[y, x]$. Observe that the F-derivative $f'(x)$ exists and is equal to $2[Bx - y, \cdot]$. Thus, $f'(x) = 0$ implies that $Bx = y$. Differentiating again, we get $f''(x) = 2B$. Thus, f is twice F-differentiable.

We next show that if

$$\theta = \lambda \sup_s \int_0^1 |k(s, t)| \, dt < 1,$$

then $f''(x)$ is bounded below. We have that

$$[f''(x) h, h] = 2[h, (I - \lambda A) h]$$

$$\geq 2(1 - \lambda \|A\|) \|h\|^2 \geq 2(1 - \theta) \|h\|^2.$$

(See Chapter I, Section C-4 and Chapter III, Section B-12, p. 129.)

Theorem

Given x_0 and y arbitrarily in $C[0, 1]$, let A, B, and K be defined as above and assume

$$\lambda \int_0^1 k(s, t)\, dt < 1, \qquad \text{for all } s \text{ in } [0,1]$$

Form the sequence $\{x_{k+1}\} = \{x_k - \gamma_k \phi(x_k)\}$, where $\phi(x_k) = 2Bx_k - y$. Here $\gamma_k \in [\delta, 2\gamma_0 - \delta]$, with

$$\frac{1}{2\gamma_0} = 1 + \lambda \sup_s \int_0^1 |k(s, t)|\, dt \quad \text{and} \quad 0 < \delta \le \gamma_0.$$

Set $z_n = y + \lambda A x_n$. Then the sequence $\{z_n\}$ converges strongly (in the uniform norm of $C[0, 1]$) to a unique point z for which $Bz = y$.

Proof

By the proof (c) of Theorem 1, Section C-5, we find that $\{x_k\}$ is a Cauchy sequence in the norm

$$[x, x]^{1/2} = \|x\|_2 = \left[\int_0^1 x^2(t)\, dt\right]^{1/2}.$$

We may not conclude that $\{x_k\}$ converges, however, because we have not established that $C[0, 1]$ is complete in the norm $\|\cdot\|$. By Section D-1, z exists. Since $-z = -y - \lambda Az$,

$$|z_n(s) - z(s)| = \lambda \int_0^1 k(s, t)(x_n(t) - z(t))\, dt$$

$$\le \lambda \left[\int_0^1 k^2(s, t)\, dt\right]^{1/2} \left[\int_0^1 (x_n(t) - z(t))^2\, dt\right]^{1/2}.$$

Also

$$\|Bx_n - y\|_2 = \|Bx_n - y - (Bz - y)\|_2 = \|B(x_n - z)\|_2.$$

Because $\{\|f'(x_n)\|\} \to 0$, by the proof of (c) it follows that

$$\{\|B(x_n - z)\|_2\} \to 0.$$

Thus, $\|x_n - z\|_2 \le \|B^{-1}\|_2 \|B(x_n - z)\|_2$, showing that, given ε, there exists N such that $n \ge N$ implies $\max\{|z_n(s) - z(s)| : s \in [0, 1]\} < \varepsilon$, which concludes the proof.

D-3. Nonlinear Integral Equations

Consider the following integral equation:

$$y(s) = x(s) - \lambda \int_0^1 g(s, t, x(t)) \, dt. \tag{I}$$

Here we are given $y \in C[0, 1]$ and g is continuous on the rectangle

$$R = \{(s, t, u): 0 \le s \le 1, 0 \le t \le 1, |u| \le \infty\}.$$

We wish to find x.

Lemma

Assume that g satisfies $|g(s, t, u) - g(s, t, u')| \le K|u - u'|$, with $K\lambda < 1$ for all $(s, t, u) \in R$. Then Equation (I) has a unique solution.

Proof

Let

$$f(x(s)) = y(s) + \lambda \int_0^1 g(s, t, x(t)) \, dt.$$

If $f(x) = x$, then x satisfies Equation (I). We calculate that $\|f(x) - f(z)\| \le \lambda K\|x - z\|$, showing that f is a contraction on $C[0, 1]$.

EXERCISE

Prove the following:

1. If $R = \{(s, t, u): 0 \le s \le 1, 0 \le t \le 1, |u| \le h\}$, $(s, t, u) \to \phi(s, t, u)$ is defined and continuous on R, and there exists $K \ge 0$ such that $|\phi(s, t, u) - \phi(s, t, u')| \le K|u' - u|$, $M = \max\{|\phi(s, t, u)|: (s, t, u) \in R\}$, and $\lambda < \min[hM^{-1}, K^{-1}]$, then the equation $x(s) = \lambda \int_0^1 \phi(s, t, x(t)) \, dt$ has a unique solution.

2. Assume that k and ϕ are continuous. The integral equation

$$f(x) = \lambda \int_a^x k(x, y) f(y) \, dy + \phi(x),$$

called *Volterra's equation*, has a unique solution of all values of λ. *Hint:* Prove that all powers of the mapping

$$f \to \lambda \int_a^x k(x, y) f(y) \, dy + \phi(x)$$

beyond a certain integer are a contraction. Recall the formula for the nth iterated integral:

$$\int_0^x du \int_0^u du \cdots \int_0^u f(t)\, dt = \frac{1}{(n-1)!} \int_0^x f(u)(x-u)^{n-1}\, du.$$

We next give an example of Equation (I), where Newton's method is applicable. Since Newton's method is an algorithm that replaces a nonlinear equation by a sequence of linear equations, success in applying this method to integral equations depends on the availability of the resolvent kernel at each iteration. If the modified Newton's method is employed, however, the resolvent kernel need not be known at every step, but necessarily at the first step. Of course it is usually not easy to find a point x_0 where the conditions of Kantorovich are satisfied. But if such a point is found, existence of a solution is guaranteed.

To apply the Newton process to Equation (I), we write

$$f(x(s)) = x(s) - \lambda \int_0^1 g(s, t, x(t))\, dt - y(s).$$

The Newtonian iteration will generate a sequence of functions in $C[0, 1]$ according to the formula

$$x_{n+1} = x_n - f'_{-1}(x_n)f(x_n).$$

The mapping $f'(x)$ is given by $I - \lambda A$, where

$$Ah = \int_0^1 g'(s, t, x(t))h(t)\, dt.$$

Therefore, $f'_{-1}(x) = (I - \lambda A)^{-1} = I + \lambda B$, where B is the resolvent kernel.

Example

$y(s) = 1 - 0.485s + s^2$, $\lambda = 1$, and $g(s, t, x(t)) = st \arctan x(t)$. The solution of Equation (I) for this example is $\bar{x}(s) = 1 + s^2$. It will be shown that if $x_0(t) = \frac{3}{2}$, the conditions of Kantorovich, Section C-4, are satisfied. We first calculate the resolvent kernel for the kernel

$$g'(s, t, x_0(t)) = \frac{(st)}{(1 + \frac{9}{4})} = \alpha st.$$

Using the Equations (IV) for the resolvent kernel (Section D-1 p. 156), we get

$$r(s, t) = \alpha st + \lambda \int_0^1 \alpha s u r(u, t)\, du, \tag{IIa}$$

$$r(s, t) = \alpha st + \lambda \int_0^1 r(s, u)\alpha t u\, du. \tag{IIb}$$

By (IIa), $r(s, t)/s$ is a function of t only; and by (IIb), $r(s, t)/t$ is a function of s only. It follows that $r(s, t)/st$ is a constant. (Why?) Hence,

$$r(s, t) = Bst.$$

Using Equation (IIa) and $\alpha = \frac{4}{13}$, we calculate that $B = \frac{12}{35}$. Thus,

$$\|f'_{-1}(z_0)\| \leq 1 + \max_s \int_0^1 |r(s, t)|\, dt = \frac{41}{35}.$$

Next we write

$$-f(x_0(s)) = 1 - 0.4854s + s^2 + s \int_0^1 t \arctan \tfrac{3}{2}\, dt - \tfrac{3}{2}$$

$$= -0.5 + 0.006012s + s^2,$$

whence $\|f(x_0)\| = 0.506012$. Since

$$k'(s, t, x(t)) = \frac{st}{(1 + x^2(t))},$$

we have

$$k''(s, t, x(t)) = \frac{-2x(t)st}{(1 + x^2(t))^2}.$$

Also,

$$f'(x(s))h(s) = h(s) + \int_0^1 k'(s, t, x(t))h(t)\, dt,$$

$$f''(x(s))h(s)k(s) = \int_0^1 k''(s, t, x(t))h(t)k(t)\, dt,$$

and

$$\max_s |k''(s, t, x(t))| = 3\left[\frac{4}{13}\right]^2 t < \frac{t}{3},$$

whence

$$\max \int_0^1 |k''(s, t, x(t))|\, dt < \tfrac{1}{6}.$$

Since $h = \beta\eta M$ (Section C-4),

$$\beta \leq \tfrac{41}{35}, \quad \eta = \|f'_{-1}(x_0)f(x_0)\| \leq \beta \|f(x_0)\| \leq (\tfrac{41}{35}) \cdot (0.506012) \quad \text{and} \quad M = \tfrac{1}{6},$$

clearly, $h \leq \tfrac{1}{2}$, showing the existence of a root of f in

$$\{x \in C[0, 1]: \max_t \|x(t) - \tfrac{3}{2}\| \leq \tfrac{82}{35}\}.$$

Since the above lemma holds for our example, we have proved for the second time the existence of a solution.

We next calculate $x_1 = x_0 - f'_{-1}(x_0)f(x_0)$. Collecting our results, we get

$$f'_{-1}(x_0)h(s) = h(s) + \int_0^1 \tfrac{12}{35}sth(t)\, dt.$$

Thus,

$$-f'_{-1}(x_0)f(x_0) = s^2 + 0.006012s - 0.5 + s\int_0^1 \tfrac{12}{35}[t^3 + 0.006012t^2 - 0.5t]\, dt.$$

The integral has the value

$$s\cdot\frac{12}{35}\cdot\frac{006012}{3} \le 0.0007s.$$

Thus, to four decimals,

$$x_1(s) = s^2 + 0.0067s + 1.$$

The true solution is $\bar{x}(s) = 1 + s^2$. Comparing, we have

$$x_1(s) - \bar{x}(s) = 0.0067s, \qquad \|x_1 - \bar{x}\| = 0.0067,$$

whereas

$$x_0(s) - \bar{x}(s) = 0.5 - s^2, \qquad \|x_0 - \bar{x}\| = 0.5,$$

which is a considerable improvement in accuracy.

SECTION E. AN APPLICATION TO CONTROL THEORY

E-1. Rendezvous Problem

We consider the following system of differential equations:

$$\begin{aligned}
\ddot{x} - 2\omega\dot{y} &= u_1,\\
\ddot{y} + 2\omega x - 3\omega^2 y &= u_2,
\end{aligned} \tag{I}$$

where the dots stand for derivatives with respect to time. These equations are the approximate differential equations of the relative motion of a "Ferry vehicle" closely approaching the circular orbit of a satellite P. The orbit of the satellite is a circle about the center of gravity of the earth, which we assume to be a fixed point, say θ. The satellite moves counterclockwise with an angular velocity ω, and the positive y axis coincides with the direction P-θ. The coordinate system (x, y) has its origin at P, and is thus a system of moving coordinates. The Ferry vehicle moves in the same plane as the satellite and has coordinates (x, y). The Ferry has a rocket engine, which is capable of thrusting at time t with accelerations $u_1(t)$ and $u_2(t)$ directed along the x and y coordinate axes. The terms involving ω in the differential equations arise because of the moving coordinate system. The system (I) arises when one considers the exact differential equations of motion, neglects the gravitational attraction between satellite and Ferry vehicle, and approximates the earth's gravitational attraction as a linear function.

We consider the following problem. Assume that at $t = 0$, initial conditions $x(0)$, $y(0)$, $\dot{x}(0)$, $\dot{y}(0)$ are prescribed. Then, as is known from the theory of linear differential equations, if u_1 and u_2 are continuous, there is a unique solution $x(t)$ and $y(t)$ satisfying (I) and the initial conditions. At time $t = \tau > 0$, terminal conditions $x(\tau) = y(\tau) = \dot{x}(\tau) = \dot{y}(\tau) = 0$ are prescribed. These conditions are the rendezvous conditions. They require that the Ferry vehicle and the satellite have identical position and velocity components at time $t = \tau$. In general, for a fixed pair of functions (u_1, u_2), these conditions will never be realized. We then ask: "Can rendezvous be achieved for some pair of continuous functions (u_1, u_2)?" If so, what is the totality of all such functions for which rendezvous is achieved? Pairs of continuous functions for which rendezvous is achieved will be called *admissible solutions*. The existence of admissible solutions is easily established by considering the solution of the systems of differential equations, which we shall do below. It will be seen that the set of solutions is very big, so that we shall prescribe more.

We ask that some condition of optimality hold with respect to the thrust functions u_1 and u_2. More specifically, we seek to minimize a real valued function of (u_1, u_2) over all admissible solutions. Let

$$|u(t)| = [u_1^2(t) + u_2^2(t)]^{1/2};$$

$|u(t)|$ is called the *thrust amplitude*. We consider minimizing the function

$$\int_0^\tau |u(t)|^p \, dt$$

for various values of p satisfying $1 < p < \infty$. When p is near 1, physical considerations show that the integral is a close measure of the fuel expended in rendezvous; when p is very large, the integral is close to

$$\max\{|u(t)| : t \in [0, \tau]\}^p,$$

as will be seen in the exercise below, and thus corresponds to the problem of achieving rendezvous with the smallest possible maximum thrust amplitude.

The solution of the differential Equations (I) may be written as

$$x(t) = \alpha_1(t) - \int_0^t [u_1(T)f_1(t - T) + u_2(T)g_1(t - T)]\, dT,$$

$$\dot{x}(t) = \alpha_2(t) - \int_0^t [u_1(T)f_2(t - T) + u_2(T)g_2(t - T)]\, dT,$$

$$y(t) = \alpha_3(t) - \int_0^t [u_1(T)f_3(t - T) + u_2(T)g_3(t - T)]\, dT,$$

$$\dot{y}(t) = \alpha_4(t) - \int_0^t [u_1(T)f_4(t - T) + u_2(T)g_4(t - T)]\, dT,$$

$$\alpha_1(t) = x_0 + \dot{x}_0\left(\left(\frac{4}{\omega}\right)\sin \omega t - 3t\right) + 6y_0(\omega t - \sin \omega t)$$

$$+ \left(\frac{2}{\omega}\right)\dot{y}_0(1 - \cos \omega t), \tag{II}$$

$$\alpha_2(t) = \dot{x}_0(4 \cos \omega t - 3) + 6\omega y_0(1 - \cos \omega t) + 2\dot{y}_0 \sin \omega t,$$

$$\alpha_3(t) = \left(\frac{2}{\omega}\right)\dot{x}_0(\cos \omega t - 1) + y_0(4 - 3 \cos \omega t) + \left(\frac{1}{\omega}\right)\dot{y}_0 \sin \omega t,$$

$$\alpha_4(t) = -2\dot{x}_0 \sin \omega t + 3\omega y_0 \sin \omega t + \dot{y}_0 \cos \omega t,$$

$$f_1(s) = \left(\frac{4}{\omega}\right)\sin \omega s - 3s, \qquad g_1(s) = 2\omega[1 - \cos \omega s],$$

$$f_2(s) = 4\omega s - 3, \qquad g_2(s) = 2 \sin \omega s,$$

$$f_3(s) = \left(\frac{2}{\omega}\right)[\cos \omega s - 1], \qquad g_3(s) = \left(\frac{1}{\omega}\right)\sin \omega s,$$

$$f_4(s) = -2 \sin \omega s, \qquad g_4(s) = \cos \omega s.$$

Let us now define a linear space as follows: The space $\mathscr{C}$ is the totality of pairs $x = (x_1, x_2)$, where $x_i \in C[0, \tau]$, $(i = 1, 2)$. The set $\{(f_i, g_i): 1 \le i \le 4\}$ is linearly independent in $\mathscr{C}$. Thus, $0 = \sum c_i(f_i(t), g_i(t))$ for all $t \in [0, \tau]$, only if $c_i = 0$, $1 \le i \le 4$. To prove this, it will be sufficient to prove that $\sum c_i f_i(t) \ne 0$ if $c_i \ne 0$. Observe first that we must have $C_3 = 0$ because there are no cosine terms in f_1, f_2, and f_4, and a linear combination of sine, cosine, and linear terms with nonzero coefficients cannot vanish. Next, C_2 must be 0 because the remaining terms contain no constant function. Finally, no nonzero linear combination of f_1 and f_4 can vanish because of the $3s$ term in f_1.

In $\mathscr{C}$ we define an inner product as follows: for $(x, y) \in \mathscr{C}$ set:

$$[x, y] = \int_0^\tau (x_1(t)y_1(t) + x_2(t)y_2(t)) \, dt.$$

Let $(u_1, u_2) = u$ and $A^i = (f_i(t - T)), g_i(t - T), (1 \leq i \leq 4)$. The first four equations of (II), evaluated at $t = \tau$, can be written

$$\alpha_i(\tau) = [A^i, u], \qquad (1 \leq i \leq 4).$$

(III)

The set

$$M = \{u \in \mathscr{C} : [A^i, u] = \alpha_i(\tau), (1 \leq i \leq 4)\}$$

is an affine subspace. (See Section A-1.) We will show that M is nonempty and that the projection map of Section A-5 is well defined on M.

Let

$$v = \sum_{i=1}^4 \lambda_i A^i.$$

Then if

$$\sum_{i=1}^4 \lambda_i [A^i, A^j] = \alpha_i(\tau),$$

v belongs to M. We have a system of four linear equations in the unknown $(\lambda_1, \cdots, \lambda_4)$, which has a unique solution if the matrix with components $[A^i, A^j], (1 \leq i \leq 4), (1 \leq j \leq 4)$, is nonsingular. Suppose, then, for some $c_j \neq 0, \sum c_j[A^i, A^j] \equiv 0$. Then

$$\sum_i c_i \sum_i c_j[A_i, A^j] = \left[\sum_i c_i A^i, \sum_i c_j A^j\right] = \|\sum_i c_i A^i\|^2 = 0,$$

showing that the set $\{A^i : 1 \leq i \leq 4\}$ is linearly dependent. This contradiction establishes the nonsingularity of the matrix. We now have a point v lying on M. Next we observe that $M = \{w + v : w \in M_0\}$, where M_0 is the subspace $\{w \in \mathscr{C} : [A^i, w] = 0, 1 \leq i \leq 4\}$.

Lemma

Given y arbitrarily in $\mathscr{C}$, there exists a point $y' \in M_0$ closest to y.

Proof

Let N = linear hull of $\{A^i : 1 \leq i \leq 4\}$. We first find y^0 in N closest to y. By Section A-7, Exercise 1, if y^0 satisfies $[y - y_0, x - y_0] = 0$ for all $x \in N$, then y is closest to N. Set $y_0 = \sum \mu_i A^i$. Then $[y - \sum \mu_i A^i, x - \sum \mu_j A^j] = 0$ for all $x \in M$ iff $[y - \mu_i A^i, \sum (\lambda_j - \mu_j) A^j] = 0$ for all $\lambda \in E_4$, this latter equation holding if and only if $[y - \sum \mu_i A^i, A^j] = 0$, $(1 \leq j \leq 4)$. Thus, if μ satisfies $[y, A^j] = \sum \mu_i [A^i, A^j]$, then $y_0 = \sum \mu_i A^i$ is closest in N to y.

Since $[y - y_0, A^j] = 0$, $(1 \leq j \leq 4)$, $y - y_0$ belongs to M_0. Take z arbitrarily in M_0 and set $y' = y - y_0$. Then

$$[y - y', z - y'] = [y_0, z - (y - y_0)] = 0.$$

Q.E.D.

Remark

If h' is closest in M_0 to h in $\mathscr{C}$ and $x \in M$, then $x + h'$ in M is closest to $x + h$ in $\mathscr{C}$.

EXERCISES

1. Prove Hölder's inequality for integrals on $C[a, b]$.

$$\int_a^b x(t) y(t)\, dt \leq \|x\|_p \|y\|_q,$$

where

$$\frac{1}{p} + \frac{1}{q} = 1 \quad \text{and} \quad \|x\|_p = \int_a^b |x(t)|^p\, dt^{1/p}.$$

Hint: Look at Chapter I, Section D-7.

2. If $q < p$, use the above inequality to prove that

$$\|x\|_q \leq \|x\|_p [b - a]^{(p-q)/qp}.$$

3. Show that $\lim_{p \to \infty} \|x\|_p = \|x\|_\infty$.

4. Show that if

$$\frac{1}{p} + \frac{1}{q} + \frac{1}{r} = 1,$$

then

$$\int_a^b |x(t) y(t) z(t)|\, dt \leq \|x\|_p \|y\|_q \|z\|_r.$$

5. Show that the functional $f(x) = \|x\|_p^p$ is convex on $\mathscr{C}$ and that therefore the set $S = \{x \in \mathscr{C} : f(x) \leq f(x_0)\}$ with x_0 arbitrary in $\mathscr{C}$ is convex.

E-2. *Application of Convex Programming*

We now consider the theorem of Section B-11. We wish to apply this theorem to the problem of minimizing $\int_0^T |u(t)|^p\, dt$ on the closed convex set M.

We notice first that the theorem was proved for a Hilbert space and not an inner product space. Upon closer examination it is evident that the completeness enters three times, the first time to guarantee the existence of a projection on C, the second in (iv) to get weak compactness and weak l.s.c., and the third in (v) to conclude that the Cauchy sequence has a limit. Since, however, we have shown that the projection operator for M is well defined, we can dispense with completeness and still obtain the conclusions of the parts (i) and (iii). For fixed $\sigma > 0$ set $\bar{x}(t) = |x(t)| + \sigma$.

We seek to minimize

$$f(x) = \int_0^{\bar{T}} |\bar{x}(t)|^p \, dt = \|\bar{x}\|_p^p$$

on the set $M = \{x \in \mathscr{C} : [A^i, x] = \alpha^i(\tau) : 1 \le i \le 4\}$, where

$$|x(t)| = [x_1^2(t) + x_2^2(t)]^{1/2} \qquad \text{and} \qquad 1 \le p \le 2.$$

We choose $x_0 = v$. (See Section E-1.) Thus, x_0 lies on M. Define $S = \{x \in \mathscr{C} : f(x) \le f(x_0)\}$. We show now that, on S, f is once F-differentiable and twice G-differentiable. Moreover, a value for γ_0 of Section B-11 will be found.

The G-differentials $f'(x)h$ and $[f''(x)h, k]$ exist (Exercise 1) and have the values

$$f'(x)h = p \int_0^T |\bar{x}(t)|^{p-2}[x_1(t)h_1(t) + x_2(t)h_2(t)] \, dt$$

$$[f''(x)h,k] = p(p-2) \int_0^T |\bar{x}(t)|^{p-4}(x_1(t)h_1(t))$$

$$+ x_2(t)h_2(t))(x_1(t)k_1(t) + x_2(t)k_2(t)) \, dt$$

$$+ p \int_0^T |\bar{x}(t)|^{p-2}(k_1(t)h_1(t) + k_2(t)h_2(t)) \, dt$$

$$= p(p-2) \int_0^T |\bar{x}(t)|^{p-2} \left(\frac{x_1(t)}{|\bar{x}(t)|} h_1(t) + \frac{x_2(t)}{|\bar{x}(t)|} h_2(t) \right) \left(\frac{x_1(t)}{|\bar{x}(t)|} k_1(t) \right.$$

$$\left. + \frac{x_2(t)}{|\bar{x}(t)|} k_2(t) \right) dt + p \int_0^{\bar{T}} |\bar{x}(t)|^{p-2}(k_1(t)h_1(t) + k_2(t)h_2(t)) \, dt.$$

Since $x_i(t) \le |x(t)|$,

$$|h_1(t)| + |h_2(t)| \le \sqrt{2}(h_1^2(t) + h_2^2(t))^{1/2} = \sqrt{2}|h(t)|,$$

and

$$h_1(t)k_1(t) + h_2(t)k_2(t) \le [h_1^2(t) + h_2^2(t)]^{1/2}[k_1^2(t) + k_2^2(t)]^{1/2} = |h(t)| \, |k(t)|,$$

whence

$$[f''(x)h, k] \leq 2p(p-2) \int_0^T |\bar{x}(t)|^{p-2}|h(t)|\,|k(t)|\,dt$$

$$+ p \int_0^{\overline{T}} |\bar{x}(t)|^{p-2}|h(t)|\,|k(t)|\,dt$$

$$= (2p^2 - 3p) \int_0^T |\bar{x}(t)|^{p-2}|h(t)|\,|k(t)|\,dt, \text{ if } p \leq 2.$$

In general, if $1 \leq p \leq \infty$

$$[f''(x)h, k] \leq p(2p+1) \int_0^T |x(t)|^{p-2}|h(t)||k(t)|dt.$$

Whence, if $1 \leq p \leq 2$

$$[f''(x)h, k] \leq \sigma^{p-2} T ||h||_2 ||k||_2.$$

We may therefore set:

$$\frac{1}{\gamma_0} = p(2p+1)\,\sigma^{p-2}T.$$

It now follows by Taylor's theorem that if $\delta = \gamma_0 \varepsilon$, we verify that f' is F-differentiable on S. Moreover, $f''(x, h, h) \geq 0$. We have thus satisfied the hypotheses (i) and (iii) of Section B-11.

EXERCISES

Prove that f' and f'' are G-differentiable. *Hint:* Use the mean-value theorem on the integrand pointwise and then use Arzela's theorem on bounded convergence[8] or the Lebesgue dominated convergence theorem.[9] Why is the number σ introduced? Given $\varepsilon > 0$, show that σ can be chosen so that if $z \in M$ satisfies $f(z) \leq \inf\{f(x): x \in M\} + \dfrac{\varepsilon}{2}$, then $||z||_p^p \leq \inf\{||x||_p^p: x \in M\} + \varepsilon$.

---

[8] Apostol, *op. cit.*, p. 405.
[9] Rudin, *op. cit.*, p. 246.

NOTES AND BIBLIOGRAPHIC MATERIAL

GENERAL COMMENTS

Much of the "nonconstructive" material in this book is taken from the common stock of knowledge in analysis, functional analysis, and convexity theory. Most of the material from functional analysis was new in the 1930s. The material on algorithms will be often cited with a reference. The reader is referred to the following books for further information on functional analysis and convexity theory.

Dunford, N., and J. Schwartz. *Linear Operators*, pt. I, Interscience Publishers, N.Y. (1958).
Eggleston, H. G. *Convexity*, Cambridge, London (1958).
Kantorovich, L. V., and G. P. Akilov. *Functional Analysis in Normed Spaces*, Macmillan, N.Y. (1964).
Valentine, F. A. *Convex Sets*, McGraw-Hill (1964).

CHAPTER I

The simple theorem on fixed points of subcontractors, p. 17, may be found in [3]. The ideas of the gradient method and steepest descent were probably first stated by Cauchy [1] and explicitly worked out in [9]. The results in D-3, p. 31, may be found in [10]. Theorem 3, p. 34, was stated in [6]. However, I do not understand the proof given there. The theorem of D-4, p. 37, may be found in [11], set in real Hilbert space. In the application of D-4, it is often advantageous not to compute second derivatives, [17].

In the following, let δ and r be positive numbers with $\delta < \frac{1}{2}$. Let f be a real-valued function defined on E_n, x^0 be an arbitrary point of E_n, and I_i be the ith column of the $n \times n$ identity matrix, I. Let S denote the level set of f at x^0, viz.: $S = [x \varepsilon E_n : f(x) \leq f(x^0)]$. Assume that for some open convex set $\hat{S}$ containing S, $f \varepsilon C^2(\hat{S})$. Let $H(x)$ denote the Hessian of f at x. Assume that for all $u \varepsilon E_n$ and for all $x \varepsilon S$, there exists a constant $\omega > 0$ such that $[u, H(x)u] \geq \omega \|u\|^2$.

An algorithm for minimizing $f(x)$ consists of performing the following computations for $k = 0, 1, 2, \cdots$:

1. Compute the $n \times n$ matrix $Q(x^k)$ whose jth column is

$$\frac{\nabla f(x^k + \Theta_k I_j) - \nabla f(x^k)}{\Theta_k}$$

where $\Theta_0 = r$ and

$$\Theta_k = r \|\phi(x^{k-1})\| \qquad \text{for } k = 1, 2, 3, \cdots,$$

in which ϕ is defined as follows:
   (a) If $k = 0$, or if $Q(x^k)$ is singular, or if

$$[\nabla f(x^k), Q^{-1}(x^k) \nabla f(x^k)] \le 0,$$

   set $\phi(x^k) = \nabla f(x^k)$.
   (b) Otherwise, set $\phi(x^k) = Q^{-1}(x^k) \nabla f(x^k)$.
2. Consider the function

$$\gamma \to g(x^k, \gamma) = \frac{f(x^k) - f(x^k - \gamma \phi(x^k))}{\gamma [\nabla f(x^k), \phi(x^k)]},$$

   If $g(x^k, 1) < \delta$, choose γ_k so that $\delta \le g(x^k, \gamma_k) \le 1 - \delta$; otherwise set $\gamma_k = 1$.
   3. Set $x^{k+1} = x^k - \gamma_k \phi(x^k)$.

Theorem

Under the assumptions stated above,

1. the sequence $[x^k]$ converges to a point z minimizing f,
2. there exists a number N such that if $k > N$ then $\gamma_k = 1$, and
3. the rate of convergence of $[x^k]$ is superlinear.

The idea of D-6, p. 41, is also from Cauchy [1] and was explicitly worked out in [9]. An application to linear approximation is given in [9]. See also Cheney [2].

CHAPTER II

The essential features of computational problems on polytopes was apparently first understood by Fourier [8] in 1831. In the early 1950s Dantzig and others (including Kantorovich in the U.S.S.R.) developed techniques for handling these problems in connection with quantitative economic questions.

The material on penalty functions, p. 59, is taken from [18]. Observe that the technique of A-2 also has a natural application to eigenvalue problems.

We now elaborate on the Remark of p. 67 by describing how the value of the support mapping is calculated in certain optimal control problems. The following is a rough sketch of the ideas.

Consider the system of linear differential equations

$$\dot{x}(t) = A(t)x(t) + B(t)u(t) \tag{I}$$

$A(t)$ $n \times n$, $B(t)$ $n \times r$, $x(t)$ $n \times 1$, and $u(t)$ $r \times 1$.

Here $r \leq n$, A and B are continuous, and each component u_j, $1 \leq j \leq r$ is measurable on $[0, T]$. Let $\|\cdot\|$ denote the Euclidean norm in E_r and assume that u satisfies:

$$\|u(t)\| \leq 1,^* \qquad \text{almost everywhere on } [0, T]. \tag{II}$$

We seek r-tuples of functions u called *admissible controls*, satisfying (II) for which $x(t)$ satisfies (I) and takes on given boundary values at $t = 0$ and $t = T$, namely:

$$x(0) \text{ at } t = 0 \qquad \text{and} \qquad x(T) = 0 \text{ at } t = T. \tag{III}$$

(Of course such admissible controls do not necessarily exist for a given initial condition $x(0)$.)

We wish to minimize $\int_0^T \|u(t)\| \, dt$ over all admissible controls for which (I) and (III) are satisfied. This problem subsumes the rendezvous problem of Section E-1 when the constraint that u is admissible is added.

Let $\overline{X}$ denote the fundamental matrix of (I), that is, $\dot{\overline{X}}(t) = A(t)\overline{X}(t)$ and $\overline{X}(0) = I$. It is known that $\overline{X}^{-1}(t)$ exists for each t. The solution of (I) can then be written:

$$x(t) = \overline{X}(t)\left[x(0) + \int_0^T \overline{X}^{-1}(s)B(s)u(s) \, ds\right].$$

Thus,

$$x(T) = 0 \Rightarrow x(0) = -\int_0^T \overline{X}^{-1}(s)B(s)u(s) \, ds = \xi(u).$$

The set $D = [\xi(u) : u \text{ is admissible}]$ is a compact subset of E_n, each point of which is an initial condition for which there exists an admissible control u for which the solution of (I) with this initial condition satisfies (III). Set:

$$\xi_{n+1}(u) = \int_0^T \|u(t)\| \, dt \qquad \text{and}$$

$$K = [(\xi(u), \xi_{n+1}(u)) : u \text{ is admissible}].$$

---

* Here 1 can be replaced by any constant. It would be absorbed in B.

For given $x \varepsilon D$, set:

$$U_x = [u : u \text{ is admissible and } \xi(u) = x] \qquad \text{and}$$

$$f(x) = \inf[\xi_{n+1}(u) : u \varepsilon U_x].$$

Clearly, $x \varepsilon D \Rightarrow -x \varepsilon D$, $f(x) \geq 0$ and $f(x) = f(-x)$.

It is known that D is convex and f is strictly convex and continuous.* More-over, for each $x \varepsilon D$ there exists $\hat{u} \varepsilon U_x$ such that $f(x) = \xi_{n+1}(\hat{u})$. See Neustadt and Paiewonsky [24]. For a given x, $f(x)$ is the minimum fuel required to achieve "rendezvous" (condition III). Since $\xi_{n+1}(u) \leq \int_0^T 1 \cdot dt = T$ for all admissible u, the set K can be written:

$$K = \{\bar{x} = (x, x_{n+1}) : x \varepsilon D \qquad \text{and} \qquad f(x) \leq x_{n+1} \leq T\}.$$

Suppose $\bar{x}$ is a boundary point of K with $x_{n+1} < T$. Then the normal λ for any supporting hyperplane for K at $\bar{x}$ has a negative $(n+1)$st component. (See p. 89.) We can normalize this last component so that it is -1. Thus, $\bar{\lambda} = (\lambda, -1)$ with $\lambda \varepsilon E_n$.

We now show how the support mapping for K can be calculated. Given any normal to K of the form $(\lambda, -1)$ we wish to maximize $[\lambda, x] - x_{n+1}$ for all $\bar{x} \varepsilon K$. Here $[,]$ is the inner product in E_n or E_r. Stated otherwise, we wish to maximize the integral

$$-\int_0^T (\lambda^* \bar{X}^{-1}(s) B(s) u(s) - \|u(s)\|) \, ds.$$

Set $-\lambda^* \bar{X}^{-1}(s) B(s) = a^*(s).$† We shall maximize the integrand $[a(s), u(s)] - \|u(s)\|$ pointwise by choosing:

$$\hat{u}(s) = \frac{a(s)}{\|a(s)\|}, \qquad \text{if } \|a(s)\| > 1 \tag{a}$$

$$\hat{u}(s) = 0 \qquad \text{if } \|a(s)\| \leq 1 \tag{b}$$

This is called a "bang-bang" control because it is switched on and off when $\|a(s)\|$ passes through the value 1. When the control is "on" it is on with full amplitude 1. For the rendezvous problem of Section E-1, it has been shown that the optimal control is "on" at most 6 times [19].

In case (a), no vector $u(s)$, u admissible, can give larger values of the integrand. In case (b), if $\|u(s)\| < 1$ and $u(s) \neq 0$, the integrand is negative.

---

* For each $\theta \varepsilon E_n$, set $a^*(t) = -\theta^* \bar{X}^{-1}(t) B(t)$. We assume that the set $[t : a(t) = 0]$ has zero measure for every $\theta \neq 0$.

† We assume the set $[t : \|a(t)\| = 1]$ has measure 0. This renders $\hat{u}$ unique. The hypothe-ses $[t : \|a(t)\| = 0]$ has measure 0 guarantees that D has interior.

Plugging in the control $\hat{u}$ defined in (a) and (b) above, we get the value of the support mapping:

$$\bar{x}(\lambda) = (x(\lambda), x_{n+1}(\lambda)) = \left(-\int_0^T \overline{X}^{-1}(s)B(s)\hat{u}(s)\,ds, \int_0^T \|\hat{u}(s)\|\,ds \right).$$

Given an initial condition $x(0)$, we should like to find a normal $(\lambda, -1)$ and the corresponding control $\hat{u}$, so that

$$x(0) = -\int_0^T \overline{X}^{-1}(s)B(s)\hat{u}(s)\,ds = x(\lambda).$$

Suppose C is a compact strictly convex subject of E_{n+1} with support function φ and that $\bar{x}^0$ is a given boundary point of C. Consider the problem of finding $\bar{\lambda}^0 \varepsilon E_{n+1}$ so that $\bar{x}(\bar{\lambda}^0) = \bar{x}^0$ where $\bar{\lambda} \rightarrow \bar{x}(\bar{\lambda})$ is the support mapping for C. Set:

$$g(\bar{\lambda}) = \varphi(\bar{\lambda}) - [\bar{x}^0, \bar{\lambda}].$$

Then $\nabla g(\bar{\lambda}) = \bar{x}(\bar{\lambda}) - \bar{x}^0$ and $\nabla g(\bar{\lambda}) = 0 \Leftrightarrow \bar{x}(\bar{\lambda}) = \bar{x}^0$. If $\lambda_{n+1} = -1$, then $\nabla g(\bar{\lambda}) = 0 \Leftrightarrow x(\lambda) = x^0$. Thus, gradient techniques may be used to solve the equation $x(\lambda) = x^0$. See p. 67, Exercise, also [12] and [24].

Much of the material in B-1 to B-3 may be found in [4]. On p. 77, we mentioned that no realistic bounds were known for the number of cycles required for the general case of the algorithm of Section B-2. In a recent paper Klee [23] has, for certain cases, given the maximum number of sets with property P (p. 79), which are subsets of a set H in E_n satisfying the Haar condition. This number, of course, is a bound for the number of cycles required by the algorithm. Suppose H contains $m > n$ points. Let $\mathscr{H}$ denote the totality of all such sets H. Let $m(n, m)$ and $M(n, m)$ denote the minimum and maximum number, respectively, of subsets with property P as H ranges over $\mathscr{H}$. Then,

$$m(n, m) = m - n \qquad \text{and}$$

$$M(n, m) \geq \binom{m - \dfrac{m-n}{2}}{n+1} + \binom{m - \dfrac{m-n+1}{2}}{n+1},$$

with equality if $n \leq 2$ or $m \leq n + 9$ or $m \leq n + 3 + 2(2n + 3)^{1/2}$. Klee conjectures that, in fact, equality always obtains.

In the convex programming algorithm, the set

$$[A_1, \cdots, A_k, A_{k+1}, \cdots, A_m] = H$$

is assumed to satisfy the Haar condition. We require subsets of H with property P which contain at least one member of $[A_1, \cdots, A_k]$. Let $m^k(n, m)$

and $M^k(n, m)$ denote the minimum and maximum of such sets as H ranges over $\mathcal{H}$. Then,

$$m^k(n, m) = \begin{cases} m - n & \text{for } m - n \leq k \leq m \\ k & \text{for } 1 \leq k \leq m - n \end{cases}$$

$$M^1(n, m) = M(n - 1, m - 1),$$

$$M^k(n, m) = M(n, m) \qquad \text{for } k \geq m - n.$$

The results of C-1 are due to Descloux [7] and those of C-2 may be found in [4].

CHAPTER III

The results of A-5 to A-7 may be found in [3]. An interesting application of the solution of systems of linear inequalities to rational approximation may be found in [13]. The theorem of B-10 may be regarded as a simple alternative to Lagrange multipliers. The material of C-4 and D-1 to D-3 follows Kantorovich in [21] and [22]. The material in C-5 is from [15]. The results of E-1 and E-2 and further results in this direction may be found in [15], [16], and [20]. The extremal of $\int_0^T \|u(t)\| \, dt$ on M is not achieved even when $\int_0^T \|u(t)\| \, dt$ is the norm in an appropriate B-space. See [5] and [19]. It is, however, achieved under the constraint that $\|u(t)\| \leq K$, a.e. on $[0, T]$ [19].

BIBLIOGRAPHY

1. Cauchy, A. Méthode générale pour la résolution des systèmes d'equations simultanées. *Comptes Rendus* 25, 2, 536 (1847).

2. Cheney, E. W. *Introduction to Approximation Theory*. McGraw-Hill, New York (1966).

3. Cheney, E. W., and A. A. Goldstein. Proximity maps for convex sets. *Proc. Amer. Math. Soc.* 6, vol. 10, no. 3, pp. 448–450 (1959).

4. —————— Newton's method for convex programming and Tchebycheff approximation. *Num. Math.* 1, 253–268 (1959).

5. —————— Tchebycheff approximation and related extremal problems. *Jour. Math. Mech.*, pp. 87–98 (see p. 90) (1965).

6. Curry, H. The method of steepest descent for non-linear minimization problems. *Quart. App. Math.* 2, 258 (1944).

7. Descloux, J. Note on convex programming. *J. Soc. Ind. Appl. Math.* 11, 737–747 (1963).

8. Fourier, J. J.-B. Solution d'une question particulière du calcul des inégalites, second extrait, *Histoire de l'Académie des Sciences*, p. 48 (1824). Also available in *Oeuvres de Fourier II*, Paris, pp. 325–328 (1890).

9. Goldstein, A. A. Cauchy's method of minimization. *Num. Math.* 4, 146–150 (1962).

10. —————— On steepest descent. *J. SIAM control Ser. A*, vol. 3, no. 1, pp. 147–151 (1965).

11. —————— On Newton's method. *Num. Math.* 7, 391–393 (1965).

12. —————— Convex programming and optimal control. *J. SIAM control Ser. A*, vol. 3, no. 1, pp. 142–146 (1965).

13. —————— On the stability of rational approximation. *Num. Math.* 5, 431–438 (1963).

14. —————— Convex programming in Hilbert space. *Bull. AMS*, vol. 70, no. 5, pp. 709–710 (Sept. 1964).

15. —————— Minimizing functionals on normed linear spaces. *J. SIAM control*, vol. 4, no. 2, pp. 81–89 (1966).

16. —————— Minimizing functionals on Hilbert space. *Computing Methods in Optimization Problems*. Academic Press, New York, pp. 159–166 (1964).

17. —————— and J. Price. *An Effective Algorithm for Minimization*. Boeing Scientific Research Laboratories, D 1-82-0566 (Sept. 1966).

18. —————— and B. R. Kripke. Mathematical programming by minimizing differentiable functions. *Num. Math.* 6, 47–48 (1964).

19. —————— and T. I. Seidman. Fuel optimal controls. *Proceedings Oberwolfach Himmelsmechanik*, Mar. 1964. Mannheim Bibliographisches Institut, pp. 257–266 (1966).

20. Johnson, F. Minimizing norms on affine subspaces. Ph.D. thesis, Univ. of Wash. (Aug. 1966).

21. Kantorovich, L. V. Functional analysis and applied mathematics (Russian). *Uspekhi Math. Nauk*, vol. III, no. 6, pp. 89–185 (1948).

22. Kantorovich, L. V., and G. P. Akilov. *Functional Analysis in Normed Spaces*. Macmillan, New York (1964).

23. Klee, V. L. A comparison of primal and dual methods for linear programming. *Num. Math.* 9, 227–235 (1966).

24. Neustadt, L. W., and B. Paiewonsky. On synthesizing optimal controls. *Proceedings Second IFAC Congress*, Butterworth, London (1965).

INDEX